*Business
Mathematics*

Brief Version

Business Mathematics

SECOND EDITION

Richard N. Aufmann
Palomar College

Vernon C. Barker
Palomar College

Joanne S. Lockwood
Plymouth State College

Business Consultant:
Diane Coleman
Department of Business
Hutchinson Community College

HOUGHTON MIFFLIN COMPANY   Boston   Toronto
Geneva, Illinois   Palo Alto   Princeton, New Jersey

Sponsoring Editor: Maureen O'Connor
Manager, Basic Books: Sue Warne
Senior Project Editor: Jean Andon
Senior Production/Design Coordinator: Pat Mahtani
Senior Manufacturing Coordinator: Marie Barnes
Marketing Manager: Mike Ginley

Cover photo: Rick Alexander and Associates, Inc.
Cover design: Catherine Hawkes

Printed in the U.S.A.

ISBN Numbers:
Student Text: 0-395-67532-4
Examination copy: 0-395-69292-X
Instructor's Manual with Solutions: 0-395-67533-2
Student Solutions Manual: 0-395-67534-0
Test Bank: 0-395-67535-9

123456789-B-96 95 94 93

Contents

3   *Decimals and Business Applications*   *69*

4   *Banking*   *105*

Preface

Business Mathematics, Second Edition, Brief Version provides a unified and comprehensive introduction to those math skills needed for success in business. Chapter 1 provides a thorough review of operations on whole numbers. Chapters 2 and 3 present operations with fractions and with decimals. Chapter 4 is devoted to topics related to banking. Chapter 5 develops the skills related to solving equations and using formulas to solve problems. Chapter 6 introduces percent. Chapters 7–10 develop concepts related to purchasing, pricing, payroll, and simple and compound interest. These chapters give the student a sound background in the application of business math concepts to the world of business today.

Interactive Approach

We believe that students learn best when they practice a skill while the concept is still fresh in their minds. Therefore, we ask students to practice early and often, learning one step at a time. Students "do" math, rather than just read about it.

We have structured each chapter of this book around this principle. Chapters are divided into objectives, so students see how all the material is related to the skills they are learning. All the pedagogical elements of the text are tied to the learning objectives (as are the test items in the Testbank). Within each objective, we follow the same step-by-step learning process:

Step 1   Students are introduced to a concept through simple and clear explanations, with plenty of examples, in the text.

Step 2   Students are then given paired examples (called "You Try It" boxes), with one example worked out for students to study and another for students to do themselves. (An example of a "You Try It" box is shown on the inside front cover of this book.) Students are able to apply what they have seen in the text and the worked example while the concept is still fresh in their minds. For any problem more complicated than simple arithmetic, the student is asked to write out a strategy for solving the problem.

Step 3   *Complete* solutions to the "You Try It" examples are given in the back of the text so students can compare their work to the solution.

Step 4   Students can then progress to the exercises at the end of the section to test their mastery. These exercises are grouped by objective to reinforce the relation between the explanation of concepts and the practice of skills. Answers to the odd-numbered exercises appear at the end of the book.

The Review/Test at the end of each chapter offers further practice. Answers to these tests are at the back of the book, and the answers are keyed to the objectives.

We believe that the immediate feedback provided to the student by this process will reinforce learning and help students retain skills faster.

Additional Learning Aids

In addition to our interactive approach, *Business Mathematics, Brief Version* provides many other features to help students not only learn basic math skills but also see how they are applied in the business world. In this edition are *Business Case Studies*. Each chapter contains one or more case studies describing a real-world business situation. Students use the skills they have learned to answer questions about the situation. These cases can be used for individual assignments or cooperative learning exercises like small group projects and class discussions.

The second edition of *Business Mathematics, Brief Version* also contains these features:

Chapter Introductions: Each chapter begins with an introduction that explains what skills will be covered in the text and how they are used in the business world.

Plentiful examples: All concepts are illustrated with real-world examples, so students always have a sense of why they are learning concepts.

Chapter Summaries: Each chapter closes with a summary that includes all key terms and essential rules covered in the chapter. The terms and rules are keyed to the chapter by objective number.

Calculator Procedures: Every chapter except Chapter 2 ("Fractions and Business Applications") includes a section on using calculators. These sections introduce the operations of calculators and provide practice in using them.

Glossary: An end-of-text glossary contains all of the terms defined in the text and listed in the Chapter Summaries.

To the Student: A note to the students at the beginning of the book offers suggestions for studying in college courses in general and for using this text effectively.

Instructional Aids

Business Mathematics, Second Edition, is also available in a 19-chapter version that covers the additional topics of annuities, business and consumer loans, inventory, depreciation, taxes and insurance, internations business, investments, financial statements, and statistics. The following ancillaries, published with the complete version, can also be used with the brief version of the text.

Instructor's Annotated Edition (IAE)   The Instructor's Annotated Edition is an exact replica of the student text, except answers to all problems in the text are printed in color on the answer lines of each exercise page. Answers to the student You Try Its, which follow the exposition in each objective, are also provided in the IAE. In addition, marginal notes to the instructor provide teaching tips and address such issues as student pitfalls, points which might be emphasized, and use of the features of the textbook.

Instructor's Manual with Solutions   The Instructor's Manual with Solutions provides complete worked-out solutions for all exercise sets, business case study questions, and Review/Tests. It also includes an overview of the text with suggestions for alternate sequencing of chapters.

Student Solutions Manual   A Student Solutions Manual containing complete solutions to odd-numbered problems is available for students to purchase.

Test Bank   The Test Bank includes two forms of each of the following printed tests: chapter tests for every chapter; cumulative tests for Chapters 1–5, 6–10, 11–15, and 16–19; and final examinations covering all 19 chapters of the text. The Testing Program also includes a data base of over 1200 test items that are organized around the same hierarchy of objectives that organize the lessons of the text. The data base can be used for selecting specific questions when preparing a test using the Instructor's Computerized Test Generator, or for selecting questions to be included on an exam prepared by hand.

Computer Supplements

Instructor's Computerized Test Generator   The Instructor's Computerized Test Generator is designed to produce free-response, objective-referenced tests for each chapter of the text. It contains over 1200 items. Cumulative tests and final examinations may also be created on the test generator. The test generator is available for IBM® PC and Macintosh®.

Computer Tutor   The Computer Tutor is an interactive instructional program for student use, covering basic skills involving whole numbers, fractions, decimals, and percent. The lesson objectives of the Computer Tutor correspond exactly to all of the lesson objectives contained within these chapters. (A correlation chart comparing the Tutor to the text is found in the Instructor's Manual.) The Computer Tutor can provide review and extra practice on material in the early chapters of the text that is considered assumed knowledge and may not be covered by the instructor during the course. The tutor is available for the IBM® PC.

Acknowledgements

Special thanks go to Diane Coleman of Hutchinson Community College, who served as business consultant for this edition. In addition, the authors would like to thank the following people who reviewed this manuscript and provided many valuable suggestions:

Sheila Baiers, Kalamazoo Valley Community College
Mary Jo Boehms, Jackson State Community College
Paul Boisvert, Robert Morris College
Doug Bronson, National Institute Center, Brown Institute Campus
Patricia Dumoulin, Elgin Community College
Iris B. Fetta, Clemson University
Judith G. Gebhart, Sinclair Community College
John Hartwick, Bucks County Community College
Frances Hendrix, Rose State College
Marilynne Hudgens, Southwestern College
Jenna Johannpeter, Belleville Area Community College
Patricia Kan, Dyke College
Edward D. Laughbaum, Columbus State Community College
Anthony Lucas, Community College of Allegheny County
Gail Marco, Robert Morris College
Paul Martin, Aims Community College
David Oliver, Edison Community College
Rachel Phillips, Southeastern Business College
Catherine H. Pirri, Northern Essex Community College
Paul N. Robillard, Bristol Community College
Carl Sonntag, Pikes Peak Community College
Russell W. Southall, Laney College
Ronald Trugman, Cañada College
Henry Weiman, Bronx Community College
Jan Whiteside, American Institute of Commerce
Thomas J. Witten, Jr., Southwest Virginia Community College

Accuracy

This text is accurate. The solutions (as well as the ancillaries that accompany the text) have been prepared by the authors themselves. In addition to the authors' own exhaustive checking, the entire book has been checked by two professors. The ancillaries have also been checked. We would like to thank the following for their help in ensuring the accuracy of this book:

Ellen Casey   Massachusetts Bay Community College
Joan Van Glabek   Edison Community College (Fla.)

Patricia Newell   Edison Community College (Fla.)
Minnie Shuler   Gulf Coast Community College (Fla.)
Sharon Testone   Onandaga Community College (N.Y.)

To the Student

Know Your Instructor's Requirements

The skills you will learn in your business mathematics course will be important in your future career—no matter what career you choose. In this textbook, we have provided you with the tools to master these skills. There's no mystery to success in this course; a little hard work and attention to your instructor will pay off. Here are a few tips to help ensure your success in this class:

To do your best in this course, you must know exactly what your instructor requires. If you don't, you probably will not meet his or her expectations and are not likely to earn a good grade in the course.

Instructors ordinarily explain course requirements during the first few days of class. Course requirements may be stated in a *syllabus*, which is a printed outline of the main topics of the course, or they may be presented orally. When they are listed in a syllabus or on other printed pages, keep them in a safe place. When they are presented orally, make sure to take complete notes. In either case, understand them completely and follow them exactly.

Take Careful Notes in Class

Attending class is vital if you are to succeed in this course. Your instructor will provide not only information but also practice in the skills you are learning. Be sure to arrive on time. You are responsible for *everything* that happens in class, even if you are absent. If you must be absent from a class session:

1. Deliver due assignments to the instructor as soon as possible.
2. Contact a classmate to learn about assignments or tests announced in your absence.
3. Hand copy or photocopy notes taken by a classmate while you were absent.

You need a notebook in which to keep class notes and records about assignments and tests. Make sure to take complete and well-organized notes. Your instructor will explain text material that may be difficult for you to understand on your own and may provide important information that is not provided in the textbook. Be sure to include in your notes everything that is written on the chalkboard.

Information recorded in your notes about assignments should explain exactly what they are, how they are to be done, and when they are due. Information about tests should include exactly what text material and topics will be covered on each test and the dates on which the tests will be given.

Survey the Chapter

Before you begin reading a chapter, take a few minutes to survey it. Glancing through the chapter will give you an overview of its content and help you see how the pieces fit together as you read.

Begin by reading the chapter title. The title summarizes what the chapter is about. Next read the section headings. The section headings summarize the major topics presented in the chapter. Then read the objectives under each section heading. The objective headings describe the learning goals for that section. Keep these headings in mind as you work through the material. They provide direction as you study.

Use the Textbook to Learn the Material

For each objective studied, read very carefully all of the material from the objective statement to the boxed examples provided for that objective. As you read, note carefully the formulas and words printed in **boldface** type. It is important for you to know these formulas and the definitions of these words.

You will note that the boxed examples come in pairs. The example on the left is worked out for you; the one on the right is left for you to do. (An example of these pairs is shown on the inside front cover of this text.) After studying the example on the left, do the example on the right. Use a pencil so that you can erase mistakes. Immediately look up the answer to this example in the answer section at the back of the text. The page number on which the solution appears is printed at the lower right-hand side of the example box. If your answer is correct, continue. If your answer is incorrect, check your solution against the one given in the answer section. It may be helpful to review the worked-out example also. Determine where you made your mistakes. Do this for each pair of examples.

Next, do the problems in the exercise set that correspond to the objective just studied. An exercise set follows every section in the textbook. Exercise sets are identified by objective. For example, the exercise set for Objective 2.2C is labeled Exercise 2.2C. The answers to all the odd-numbered exercises appear in the answer section in the back of the book. Check your answers to the exercises against these.

If you have difficulty solving problems in the exercise set, review the material in the text. Many examples are solved within the text material. Review the solutions to these problems. Reread the boxed examples provided for the objective. If, after checking these sources and trying to find your mistakes, you are still unable to solve a problem correctly, make a note of the exercise number so that you can ask someone for help with that problem.

Follow this procedure for each objective assigned by your instructor.

Review Material

Reviewing material is the repetition that is essential for learning. Much of what we learn is soon forgotten unless we review it. If you find that you do not remember information that you studied previously, you probably have not reviewed it sufficiently. *You will remember best what you review most.*

One method of reviewing material is to begin a study session by reviewing a concept you have studied previously. For example, before trying to solve a new type of problem, spend a few minutes solving a kind of problem you already know how to solve. Not only will you provide yourself with the review practice you need, but you are also likely to put yourself in the right frame of mind for learning how to solve the new type of problem.

Use the End-of-Chapter Material

To help you review the material presented within a chapter, a Chapter Summary appears at the end of each chapter. In the Chapter Summary, definitions of the important terms and concepts introduced in the chapter are provided under "Key Words." Listed under "Essential Rules" are the formulas and procedures presented in the chapter. After completing a chapter, be sure to read the Chapter Summary. Use it to check your understanding of the material presented and to determine what concepts you need to review. Printed within parentheses following each Key Word or Essential Rule is the objective in which it is presented in the chapter. Return to that objective in the textbook to restudy the concept.

Each chapter ends with a Review/Test. The problems it contains summarize what you should have learned when you have finished the chapter. Do the exercises in the Review/Test as you prepare for an examination. Check your answers against those in the back of the book. Answers to every exercise in a Review/Test are provided there. The objective being tested by any particular problem is written in parentheses following the answer. For example, for the Review/Test for Chapter 3, the answer "8. 2.4723 (Objective 3.2B)" indicates that the answer to Exercise 8 in the Chapter 3 Review/Test is 2.4723 and that Exercise 8 corresponds to Objective 3.2B in Chapter 3. For any problem you answer incorrectly, review the material corresponding to that objective in the textbook. Determine *why* your answer was wrong.

Calculator pages are included in the end-of-chapter material in the text. The more quickly and efficiently you can use your calculator, the easier it will be for you to solve many of the problems in this book. Take the time to learn how your calculator functions and make use of its capabilities. Doing so will ultimately save you a great deal of time.

Finding Good Study Areas

Find a place to study where you are comfortable and can concentrate well. Many students find the campus library to be a good place. You might select two or three places at the college library where you like to study. Or there may be a small, quiet lounge on the third floor of a building where you find you can study well. Take the time to find places that promote good study habits.

Determining When to Study

Spaced practice is generally superior to massed practice. For example, four half-hour study periods will produce more learning than one two-hour study session. The following suggestions may help you decide when you will study.

1. A free period immediately before class is the best time to study about the lecture topic for the class.
2. A free period immediately after class is the best time to review notes taken during the class.
3. A brief period of time is good for reciting or reviewing information.
4. A long period of an hour or more is good for doing challenging activities such as learning to solve a new type of problem.
5. Free periods just before you go to sleep are good times for learning information. (There is evidence that information learned just before sleep is remembered longer than information learned at other times.)

Determining How Much to Study

Instructors often advise students to spend twice the amount of time outside of class studying as they spend in the classroom. For example, if a course meets for three hours each week, customarily instructors advise students to study for six hours each week outside of class.

Your business mathematics course requires the learning of skills, which are abilities acquired through practice. It is often necessary to practice a skill more than a teacher requires. For example, this textbook may provide 50 practice problems on a specific objective and the instructor may assign only 25 of them. However, some students may need to do 30, 40, or all 50 problems.

If you are an accomplished athlete, musician, or dancer, you know that long hours of practice are necessary to acquire a skill. Do not cheat yourself of the practice you need to develop the abilities taught in this course.

Study followed by reward is usually productive. Schedule something enjoyable to do following study sessions. If you know that you have only two hours to study because you have scheduled a pleasant activity for yourself, you may be inspired to make the best use of the two hours that you have set aside for studying.

Keep Up to Date with Course Work

College terms start out slowly. Then they gradually get busier and busier, reaching a peak of activity at final examination time. If you fall behind in the work for a course, you will find yourself trying to catch up at a time when you are very busy with all of your other courses. Don't fall behind—keep up to date with course work.

Keeping up with course work is doubly important for a course in which information and skills learned early in the course are needed to learn information and skills later in the course. Business mathematics is such a course. Skills must be learned immediately and reviewed often. For example, solving problems involving percents, a skill learned in Chapter 4, will be used through the remainder of the course.

Your instructor gives assignments to help you acquire a skill or understand a concept. Do each assignment as it is assigned or you may well fall behind and have great difficulty catching up. Keeping up with course work also makes it easier to prepare for each exam.

Be Prepared for Tests

The Review/Test at the end of a chapter should be used to prepare for an examination. We suggest that you try the Review/Test a few days before your actual exam. Do these exercises in a quiet place and try to complete the exercises in the same amount of time as you will be allowed for your exam. When completing the exercises, practice the strategies of successful test takers: 1) look over the entire test before you begin to solve any problem; 2) write down any rules or formulas you may need so they are readily available; 3) read the directions carefully; 4) work the problems that are easiest for you first; 5) check your work, looking particularly for careless errors.

When you have completed the exercises in the Review/Test, check your answers. If you missed a question, review the material in that objective and rework some of the exercises from that objective. This will strengthen your ability to perform the skills in that objective.

Get Help for Academic Difficulties

If you do have trouble in this course, teachers, counselors, and advisers can help. They usually know of study groups, tutors, or other sources of help that are available. They may suggest visiting an office of academic skills, a learning center, a tutorial service, or some other department or service on campus.

Students who have already taken the course and who have done well in it may be a source of assistance. If they have a good understanding of the material, they may be able to help by explaining it to you.

1

Whole Numbers and Business Applications

*O*BJECTIVES

1.1A   To write whole numbers in words and in standard form

1.1B   To round a whole number to a given place value

1.2A   To add whole numbers

1.2B   To subtract whole numbers

1.2C   To solve application problems

1.2D   To solve problems involving a quota

1.3A   To multiply whole numbers

1.3B   To divide whole numbers

1.3C   To find unit cost and total cost

*M*ATHEMATICS IS A tool. It is used to help us better understand our world and relationships that exist within it. Mathematics is used in physics, economics, engineering, meteorology, computer science, and space exploration, to name but a few areas of its application. And it is used in business.

As a consumer, you already have some knowledge of the business world. You have bought goods at clothing stores, hardware stores, bookstores, music stores, shoe stores, grocery stores, and jewelry stores. You have purchased services through doctors, dentists, hair stylists, dry cleaners, banks, and restaurants.

In the paragraph above, the word *consumer* was used. **Consumers** purchase goods and services for their personal use, rather than with the intent of reselling them. The following is a list of other business terms with which you are probably already familiar, although you may not know precise definitions. Which of these terms involve the use of numbers?

Budget	Percent
Credit	Price
Cost	Profit
Discount	Rent
Expense	Salary
Insurance	Sale
Investment	Stocks and bonds
Loan	Taxes
Loss	

The answer is that they all involve the use of numbers. And that is where we begin our study of business mathematics. In Chapter 1, the operations of addition, subtraction, multiplication, and division of whole numbers are presented, along with some applications of mathematics to business. This material is a foundation for all the remaining chapters of this textbook.

To ensure that you get off to a good start in your study of business mathematics, we recommend that you read the Student Preface. Know and understand how this textbook works for you in the learning process.

Consider reading, on a regular basis, the business section of any major newspaper. (Libraries generally have daily newspapers in the reference section.) Although you may not understand all the terms or discussions at first, the exposure will broaden your understanding. And as you continue in your study of business mathematics, you will come to understand more and more of what you read and will develop a greater appreciation for the application of your studies and for the business world itself.

SECTION 1.1   Introduction to Whole Numbers

Objective 1.1A   *To write whole numbers in words and in standard form*

The **whole numbers** are 0, 1, 2, 3, 4, 5, 6, 7, 8, 9, 10, 11, 12, 13, and so on.

When a whole number is written using the digits 0, 1, 2, 3, 4, 5, 6, 7, 8, and 9, it is said to be in **standard form.** The position of each digit in the number determines the digit's **place value.** The following diagram shows a **place-value chart** naming the first thirteen place values. The number 37,462 is in standard form and has been entered in the chart.

In the number 37,462, the position of the digit 3 determines that its place value is ten-thousands.

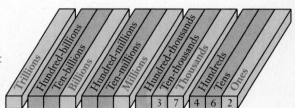

When a number is written in standard form, each group of digits separated by a comma is called a **period.** The number 3,786,451,294 has four periods. The period names are shown in color in the place-value chart above.

To write a number in words, start from the left. Name the number in each period. Then write the period name in place of the comma.

3,786,451,294 is read three billion seven hundred eighty-six million four hundred fifty-one thousand two hundred ninety-four.

To write a whole number in standard form, write the number named in each period and replace each period name with a comma.

Four million sixty-two thousand five hundred eighty-four is written 4,062,584. The zero is used as a place holder for the hundred-thousands' place.

Example 1

Write 25,478,083 in words.

Solution

twenty-five million four hundred seventy-eight thousand eighty-three

Example 2

Write three hundred three thousand three in standard form.

Solution

303,003

You Try It 1

Write 36,462,075 in words.

Your solution

You Try It 2

Write four hundred fifty-two thousand seven in standard form.

Your solution

Solutions on p. A3

Objective 1.1B   *To round a whole number to a given place value*

When the distance to the moon is given as 240,000 miles, the number represents an approximation to the true distance. Giving an approximate value for an exact number is called **rounding.** A number is always rounded to a given place value.

A whole number is rounded to a given place value by looking at the first digit to the right of the given place value.

If the digit to the right of the given place value is less than 5, that digit and all digits to the right are replaced by zeros.

Round 13,834 to the nearest hundred.

⌐——— Given place value

13,834

└——— 3 is less than 5

13,834 rounded to the nearest hundred is 13,800.

If the digit to the right of the given place value is greater than or equal to 5, increase the digit in the given place value by 1, and replace all other digits to the right by zeros.

Round 386,217 to the nearest ten-thousand.

⌐——— Given place value

386,217

└——— 6 is greater than 5

386,217 rounded to the nearest ten-thousand is 390,000.

Example 3

Round 525,453 to the nearest ten-thousand.

Solution

⌐——— Given place value (ten-thousand)

525,453

└——— 5 is equal to 5

525,453 rounded to the nearest ten-thousand is 530,000.

You Try It 3

Round 368,492 to the nearest ten-thousand.

Your solution

Example 4

Round 1972 to the nearest hundred.

Solution

⌐——— Given place value (hundred)

1972

└——— 7 is greater than 5

1972 rounded to the nearest hundred is 2000.

You Try It 4

Round 3962 to the nearest hundred.

Your solution

Solutions on p. A3

EXERCISE 1.1A

Write the number in words.

1. 805   1. _____

2. 609   2. _____

3. 485   3. _____

4. 576   4. _____

5. 2675   5. _____

6. 3790   6. _____

7. 42,928   7. _____

8. 58,473   8. _____

9. 80,106   9. _____

10. 60,930   10. _____

11. 356,943   11. _____

12. 498,512   12. _____

13. 3,697,483   13. _____

14. 6,842,715   14. _____

Write the number in standard form.

15. eighty-five   15. _____

16. three hundred fifty-seven   16. _____

17. four hundred six   17. _____

18. two thousand three hundred ninety   18. _____

19. three thousand four hundred fifty-six   19. _____

20. sixty-three thousand seven hundred eighty   20. _____

21. fifty-two thousand one hundred forty-eight   21. _____

22. seventy thousand nine hundred seventy-one   22. _____

23. six hundred nine thousand nine hundred forty-eight   23. _____

24. seven million twenty-four thousand seven hundred nine   24. _____

25. four million three thousand two

26. six million five thousand eight

27. nine million four hundred sixty-three thousand

28. sixteen billion eight hundred thirty million

EXERCISE 1.1B

Round the number to the given place value.

29.	926	tens	**30.**	845	tens
31.	1439	hundreds	**32.**	3973	hundreds
33.	7238	thousands	**34.**	7609	thousands
35.	43,607	thousands	**36.**	92,466	thousands
37.	253,678	thousands	**38.**	647,989	thousands
39.	55,211	ten-thousands	**40.**	27,359	ten-thousands
41.	926,553	ten-thousands	**42.**	382,063	ten-thousands
43.	4,279,355	millions	**44.**	9,604,235	millions
45.	5,568,929	thousands	**46.**	8,435,579	thousands
47.	6,845,733	ten-thousands	**48.**	7,384,956	ten-thousands
49.	39,875,688	thousands	**50.**	24,119,825	thousands

25. _____
26. _____
27. _____
28. _____
29. _____
30. _____
31. _____
32. _____
33. _____
34. _____
35. _____
36. _____
37. _____
38. _____
39. _____
40. _____
41. _____
42. _____
43. _____
44. _____
45. _____
46. _____
47. _____
48. _____
49. _____
50. _____

 **ECTION 1.2**   ## Addition and Subtraction of Whole Numbers

Objective 1.2A   *To add whole numbers*

Addition is the process of finding the total of two or more numbers.

The total of $3 and $4 is $7.

$$\$3 \quad + \quad \$4 \quad = \quad \$7$$

Addend + Addend = Sum

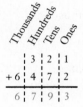

To add large numbers, begin by arranging the numbers vertically, keeping the digits of the same place value in the same column.

Add: $321 + 6472$

Add the digits in each column.

```
      Thousands Hundreds Tens Ones
          |  3 | 2 | 1
      + 6 |  4 | 7 | 2
        6 |  7 | 9 | 3
```

When the sum of the digits in a column exceeds 9, the addition involves "carrying."

Add: $487 + 369$

Add the ones' column.
$7 + 9 = 16$ (1 ten + 6 ones).
Write the 6 in the ones' column and carry the 1 ten to the tens' column.

```
   Hundreds Tens Ones
          1
      4 | 8 | 7
    + 3 | 6 | 9
          |   | 6
```

Add the tens' column.
$1 + 8 + 6 = 15$ (1 hundred + 5 tens).
Write the 5 in the tens' column and carry the 1 hundred to the hundreds' column.

```
    1 1
    4 8 7
  + 3 6 9
      5 6
```

Add the hundreds' column.
$1 + 4 + 3 = 8$ (8 hundreds).
Write the 8 in the hundreds' column.

```
    1 1
    4 8 7
  + 3 6 9
    8 5 6
```

When adding a column of numbers, you can increase your speed and accuracy by recognizing sums of 10.

In the example at the right, there are two combinations of 10: the 6 + 4 and the 3 + 7. Find these sums first. Then add the remaining digits to the total.

$$
\begin{array}{r}
6 \\
4 \\
1 \\
3 \\
7 \\
+\ 5 \\
\hline
26
\end{array}
$$
⎤ 10 (for 6, 4)
⎤ 10 (for 3, 7)

Example 1

Add: 89 + 36 + 98

Solution

$$
\begin{array}{r}
2 \\
89 \\
36 \\
+\ 98 \\
\hline
223
\end{array}
$$

You Try It 1

Add: 95 + 88 + 67

Your solution

Example 2

Add: 41,395 + 4327 + 497,625

Solution

$$
\begin{array}{r}
111\ 11 \\
41,395 \\
4,327 \\
+\ 497,625 \\
\hline
543,347
\end{array}
$$

You Try It 2

Add: 392 + 4079 + 89,035 + 4992

Your solution

Solutions on p. A3

Objective 1.2B   *To subtract whole numbers*

Subtraction is the process of finding the difference between two numbers.

The difference between $8 and $5 is $3.

   $8   – $5   = $3

Minuend – Subtrahend = Difference

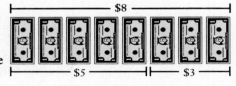

Note that addition and subtraction are related.

$$
\begin{array}{ll}
\text{Subtrahend} & \$5 \\
+\ \text{Difference} & +\ \$3 \\
\hline
=\ \text{Minuend} & \$8
\end{array}
$$

The fact that the sum of the subtrahend and the difference equals the minuend can be used to check the accuracy of subtraction.

To subtract large numbers, begin by arranging the numbers vertically, keeping the digits of the same place value in the same column. Then subtract the digits in each column.

Subtract 8955 − 2432 and check.

$$\begin{array}{r}
8\ \ 9\ \ 5\ \ 5 \\
-2\ \ 4\ \ 3\ \ 2 \\
\hline
6\ \ 5\ \ 2\ \ 3
\end{array}$$

Check:
$$\begin{array}{lr}
\text{Subtrahend} & 2432 \\
+\ \text{Difference} & +\,6523 \\
\hline
=\ \text{Minuend} & 8955
\end{array}$$

When the lower digit is larger than the upper digit, subtraction involves "borrowing."

Subtract: 692 − 378

$$\begin{array}{r}
6\ \ \overset{8+1}{\cancel{9}}\ \ 2 \\
-3\ \ 7\ \ 8
\end{array}$$

Because 8 is greater than 2, borrowing is necessary.
9 tens =
8 tens + 1 ten.

$$\begin{array}{r}
6\ \ \overset{8+\boxed{1}}{9}\ \ \overset{10}{2} \\
-3\ \ 7\ \ 8
\end{array}$$

Borrow 1 ten from the tens' column and write 10 in the ones' column.

$$\begin{array}{r}
6\ \ \overset{8}{\cancel{9}}\ \ \overset{12}{\cancel{2}} \\
-3\ \ 7\ \ 8
\end{array}$$

Add the borrowed 10 to 2.

$$\begin{array}{r}
6\ \ \overset{8}{\cancel{9}}\ \ \overset{12}{\cancel{2}} \\
-3\ \ 7\ \ 8 \\
\hline
3\ \ 1\ \ 4
\end{array}$$

Subtract the digits in each column.

Subtract 751 − 234 and check.

Step 1
$$\begin{array}{r}
7\ \overset{4}{\cancel{5}}\ \overset{11}{\cancel{1}} \\
-2\ 3\ 4 \\
\hline
\end{array}$$

Step 2
$$\begin{array}{r}
7\ \overset{4}{\cancel{5}}\ \overset{11}{\cancel{1}} \\
-2\ 3\ 4 \\
\hline
5\ 1\ 7
\end{array}$$

Check:
$$\begin{array}{r}
\overset{1}{\ }\ 234 \\
+\,517 \\
\hline
751
\end{array}$$

There may be more than one column in which borrowing is necessary.

Subtract 1234 − 485 and check.

Step 1
$$\begin{array}{r}
1\ 2\ \overset{2}{\cancel{3}}\ \overset{14}{\cancel{4}} \\
-\ \ 4\ 8\ 5 \\
\hline
9
\end{array}$$

Step 2
$$\begin{array}{r}
1\ \overset{1}{\cancel{2}}\ \overset{12}{\cancel{3}}\ \overset{14}{\cancel{4}} \\
-\ \ 4\ 8\ 5 \\
\hline
4\ 9
\end{array}$$

Step 3
$$\begin{array}{r}
\overset{0}{\cancel{1}}\ \overset{11}{\cancel{2}}\ \overset{12}{\cancel{3}}\ \overset{14}{\cancel{4}} \\
-\ \ 4\ 8\ 5 \\
\hline
7\ 4\ 9
\end{array}$$

Check:
$$\begin{array}{r}
\overset{11}{\ }\ 485 \\
+\,749 \\
\hline
1234
\end{array}$$

Subtraction with a zero in the minuend involves repeated borrowing.

Subtract: 3904 − 1775

$$\begin{array}{r} 8\ 10 \\ 3\ \cancel{9}\ \cancel{0}\ 4 \\ -1\ 7\ 7\ 5 \\ \hline \end{array}$$

$$\begin{array}{r} 9 \\ 8\ \cancel{10}\ 14 \\ 3\ \cancel{9}\ \cancel{0}\ \cancel{4} \\ -1\ 7\ 7\ 5 \\ \hline \end{array}$$

$$\begin{array}{r} 9 \\ 8\ \cancel{10}\ 14 \\ 3\ \cancel{9}\ \cancel{0}\ \cancel{4} \\ -1\ 7\ 7\ 5 \\ \hline 2\ 1\ 2\ 9 \end{array}$$

Here 5 is greater than 4, and there is a 0 in the tens' column. Borrow 1 hundred from the hundreds' column and write 10 in the tens' column.

Borrow 1 ten from the tens' column and add 10 to the 4 in the ones' column.

Subtract the digits in each column.

Note that for the preceding example, the borrowing could be performed as follows:

Borrow 1 from 90. (90 − 1 = 89. The 8 is in the hundreds' column. The 9 is in the tens' column.) Add 10 to the 4 in the ones' column. Then subtract the numbers in each column.

$$\begin{array}{r} 8\ 9\ 14 \\ 3\ \cancel{9}\ \cancel{0}\ \cancel{4} \\ -1\ 7\ 7\ 5 \\ \hline 2\ 1\ 2\ 9 \end{array}$$

Example 3

Subtract 63,221 − 23,954 and check.

Solution

$$\begin{array}{r} 12\ \ 11\ \ 11 \\ 5\ \ \cancel{2}\ \ \cancel{1}\ \ \cancel{1}\ \ 11 \\ \cancel{6}\ \ \cancel{3},\ \cancel{2}\ \ \cancel{2}\ \ \cancel{1} \\ -2\ \ 3,\ 9\ \ 5\ \ 4 \\ \hline 3\ \ 9,\ 2\ \ 6\ \ 7 \end{array}$$

Check: 23,954
       + 39,267
       ————————
         63,221

You Try It 3

Subtract 54,562 − 14,485 and check.

Your solution

Example 4

Subtract 46,005 − 32,167 and check.

Solution

$$\begin{array}{r} 5\ \ 9\ \ 9\ 15 \\ 4\ \cancel{6},\ \cancel{0}\ \cancel{0}\ \cancel{5} \\ -3\ 2,\ 1\ 6\ 7 \\ \hline 1\ 3,\ 8\ 3\ 8 \end{array}$$

There are two zeros in the minuend. Borrow 1 from 600 (600 − 1 = 599) and add 10 to the 5 in the ones' column.

Check: 32,167
       + 13,838
       ————————
         46,005

You Try It 4

Subtract 64,003 − 54,936 and check.

Your solution

Solutions on p. A3

*O*bjective 1.2C   *To solve application problems*

To solve an application problem, first read the problem carefully. The *strategy* involves identifying the quantity to be found and planning the steps necessary to find that quantity. The *solution* involves performing each operation stated in the strategy and writing the answer.

Example 5

During the first ten months of the year, 1578 computers were sold at a Radio Shack dealership. In November, 167 computers were sold at the dealership, and 203 computers were sold in December. What was the total number of computers sold during the entire year?

Strategy

To find the total number of computers sold, add the numbers sold in November and December to the number sold during the first ten months of the year (1578 + 167 + 203).

Solution

```
  1578
   167
+  203
  1948
```

A total of 1948 computers were sold during the year.

You Try It 5

A dentist treated 376 patients in January and 449 patients in February. What was the total number of patients treated by the dentist during those two months?

Your strategy

Your solution

Example 6

A customer makes a down payment of $3475 on a car costing $12,550. Find the amount that remains to be paid.

Strategy

To find the amount that remains to be paid, subtract the down payment ($3475) from the cost ($12,550).

Solution

```
 $12,550
 − 3,475
  $9,075
```

The amount that remains to be paid is $9075.

You Try It 6

Suppose the down payment on an IBM computer system that costs $3350 is $875. Find the amount that remains to be paid.

Your strategy

Your solution

Solutions on p. A3

Objective 1.2D   *To solve problems involving a quota*

A quota establishes either an upper limit or a lower limit. For example, an import quota dictates the *maximum* amount of a commodity that may be imported into a country during a given period. A government may impose a quota on imports in order to protect domestic industries from foreign competition.

A sales quota establishes a *minimum* amount of sales expected during a given period. In setting sales quotas, managers take into consideration the nature of the sales representative's territory, the salesperson's past selling experience, and expectations expressed by the sales force. Sales quotas enable a company to forecast sales and, therefore, future income, which provides a basis on which to set up company budgets.

Example 7

A sales representative has a monthly sales quota of 500 units. The representative sold 125 units during the first week, 113 units during the second week, and 147 units during the third week of the month. How many units must be sold before the end of the month if the salesperson is to meet the quota?

Strategy

To find the number of units that must be sold before the end of the month:

* Find the total number of units sold during the first three weeks of the month by adding the numbers of units sold during the first, second, and third weeks of the month (125 + 113 + 147).

* Subtract the sum from the quota (500).

Solution

```
  125
  113
+ 147
  385
```

The number of units sold during the first three weeks of the month was 385.

```
  500
− 385
  115
```

The sales representative must sell 115 units before the end of the month in order to meet the quota.

You Try It 7

A sales representative has a sales quota of 650 units for the first three months of the year. The representative sold 225 units in January and 198 units in February. How many units must be sold during March if the salesperson is to meet the quota?

Your strategy

Your solution

Solution on p. A4

*E*XERCISE 1.2A

Add.

1.	421 + 308	**2.**	8092 + 6307	
3.	71,092 + 85,407	**4.**	923,571 + 863,117	

5.	859 + 725	**6.**	1897 + 3246	
7.	36,925 + 69,392	**8.**	878 737 + 189	

9.	482 309 + 551	**10.**	9409 3253 + 7078	
11.	67,428 32,171 + 20,971	**12.**	439 332 589 + 528	

13. $2709 + 658 + 10,935$

14. $8707 + 216 + 90,714$

15. $20,958 + 3218 + 42$

16. $80,973 + 5168 + 29$

17. $392 + 37 + 10,924 + 621$

18. $694 + 62 + 70,129 + 217$

19. $692 + 2107 + 3196 + 92$

20. $294 + 1029 + 7935 + 65$

21. $97 + 7234 + 69,552 + 276$

22. $87 + 1698 + 27,317 + 727$

23. $62 + 329 + 8954 + 1072$

24. $654 + 7293 + 237 + 33$

25. $87 + 946 + 6571 + 2103$

26. $994 + 91,764 + 6571 + 2103$

1. _____
2. _____
3. _____
4. _____
5. _____
6. _____
7. _____
8. _____
9. _____
10. _____
11. _____
12. _____
13. _____
14. _____
15. _____
16. _____
17. _____
18. _____
19. _____
20. _____
21. _____
22. _____
23. _____
24. _____
25. _____
26. _____

EXERCISE 1.2B

Subtract.

27.	89 − 23	**28.**	1202 − 701	

29.   8974
    − 3972

30.   8976
    − 7463

31.   88
   − 79

32.   993
    − 537

33.   768
    − 194

34.   893
    − 874

35.   3129
    − 1785

36.   7403
    − 294

37.   2600
    − 1972

38.   8003
    − 1735

39.   470 − 92

40.   674 − 337

41.   4350 − 729

42.   7236 − 1978

43.   3700 − 58

44.   8052 − 2709

45.   70,702 − 4239

46.   10,024 − 9306

47.   12,701 − 8624

48.   14,316 − 942

49.   80,053 − 27,649

50.   95,432 − 87,857

51.   13,806 − 9439

52.   68,023 − 29,174

27. _____

28. _____

29. _____

30. _____

31. _____

32. _____

33. _____

34. _____

35. _____

36. _____

37. _____

38. _____

39. _____

40. _____

41. _____

42. _____

43. _____

44. _____

45. _____

46. _____

47. _____

48. _____

49. _____

50. _____

51. _____

52. _____

EXERCISE 1.2C

Solve.

53. Central Construction Company contracted to complete an addition to an office building. The contract specified that the work would be completed in 90 days. The construction company began the work 43 days ago. How many days are left for the construction company to complete the job?

54. A customer has an outstanding bill of $175. If the customer makes a payment of $85, how much remains to be paid by the customer?

55. A clothing manufacturer had an inventory of 3000 sports jackets before shipping an order for 550 sports jackets. Find the manufacturer's current inventory of sports jackets.

56. A manufacturer of sports equipment had an inventory of 2750 tennis rackets before filling an order for 325 tennis rackets. Find the manufacturer's current inventory of tennis rackets.

57. A car dealer offers a customer a trade-in allowance of $3500 toward the purchase of a new car selling for $18,500. How much would the customer have to pay in order to purchase the new car?

58. The down payment on a computer system costing $12,490 is $1249. Find the amount that remains to be paid.

59. At beginning of the month, the odometer of a sales representative's car read 28,479 miles. At the end of the month, the odometer read 30,362 miles. How many miles was the car driven during the month?

60. The stock clerk of the Supplies Department of a company fills two orders, one for one dozen pencils and the other for two dozen pencils. If there were 120 pencils in the supply room before the orders were filled, how many pencils remained in the supply room after the orders were filled?

61. The records for the supply room of an office show that 473 boxes of company letterhead stationery were on hand before 242 boxes were distributed to various departments throughout the company and a shipment of 500 boxes was received from the supplier. Find the number of boxes of letterhead stationery that are currently in the supply room.

62. Benton Hardware Store had 150 shovels on display and in the storage room. In one month, 73 shovels were sold and 5 shovels were returned by dissatisfied customers. How many shovels did the hardware store have at the end of the month?

63. The shoe department of Adams Department Store had an inventory of 453 pairs of shoes. During one week, 162 pairs of shoes were purchased and 18 pairs were returned to the store. Find the number of pairs of shoes in inventory at the end of the week.

53. _____

54. _____

55. _____

56. _____

57. _____

58. _____

59. _____

60. _____

61. _____

62. _____

63. _____

EXERCISE 1.2D

Solve.

64.   _____

65.   _____

66.   _____

67.   _____

68.   _____

69.   _____

70.   _____

71.   _____

72.   _____

64. In 1921 Congress passed the first legislation restricting the number of immigrants to be admitted each year to the United States. The quota, excluding North and South America, was 357,803. The quota was reduced to 150,000 in 1929. Find the difference between the 1921 quota and the 1929 quota.

65. In 1921 the immigration quota was set at 357,803. In 1952 the Immigration and Nationality Act established a new immigration quota of 158,361 per year. Find the difference between the 1921 quota and the 1952 quota.

66. Brooker Company divides its sales area into five territories. The sales quotas for the territories, expressed in dollars, are $75,000, $87,500, $68,000, $96,000, and $82,500. What are the total sales expected from the five territories?

67. The Mills Company has divided its sales area into four territories and has established the following sales quotas for the territories: $165,000, $195,000, $155,000, and $130,000. Find the total sales expected from the four territories.

68. The sales quotas for a company's four sales territories are $1,750,000, $2,250,000, $2,500,000, and $2,750,000. Find the total sales expected from the four territories.

69. Within the week, a sales representative must contact a minimum of 200 potential customers. If the representative contacted 46 people on Monday, 43 on Tuesday, and 39 on Wednesday, how many potential customers must be contacted on Thursday and Friday?

70. A sales representative sold 859 units during the first quarter of the year, 836 during the second quarter, and 924 during the third quarter. How many units must be sold during the fourth quarter if the representative's annual sales quota is 3500 units?

71. A sales representative sold 247 units in January, 293 units in February, and 208 units in March. How many units must be sold in April if the representative's four-month sales quota is 1000 units?

72. A sales representative's total sales, in dollars, during the first quarter of the year were $87,700. During the second quarter, total sales were $92,600, and during the third quarter, total sales were $79,800. What must the representative's total sales for the fourth quarter be if the representative's annual sales quota is $350,000?

 **Multiplication and Division of Whole Numbers**

*O*bjective 1.3A　*To multiply whole numbers*

Six boxes of speakerphones are ordered. Each box contains 8 speakerphones. How many speakerphones are ordered?

This problem can be worked by adding six 8's.

$8 + 8 + 8 + 8 + 8 + 8 = 48$

This problem involves repeated addition of the same number and can be worked by a shorter process called **multiplication**.

Multiplication is the repeated addition of the same number.

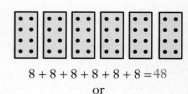

$8 + 8 + 8 + 8 + 8 + 8 = 48$
or

The numbers that are multiplied are called **factors**. The answer is called the **product**.

$6　\times　8　=　48$
Factor × Factor = Product

The times sign, ×, is one symbol that is used to mean multiplication. Another symbol commonly used for this purpose is a dot placed between the numbers. Parentheses also indicate multiplication.

$7 \times 8 = 56$　　　$7 \cdot 8 = 56$　　　$7(8) = 56$

Multiplication of large numbers requires the repeated use of the basic facts for multiplying one-digit numbers.

Multiply: 37×4

Multiply 4×7.
$4 \times 7 = 28$ (2 tens $+ 8$ ones)
Write the 8 in the ones' column and carry the 2 to the tens' column.

$$\begin{array}{r} 2 \\ 3\,|7 \\ \times\ |4 \\ \hline |8 \end{array}$$

The 3 in 37 is 3 tens.
Multiply 4×3 tens.
　　　　4×3 tens $= 12$ tens
Add the carry digit.　$\underline{+\ 2}$ tens
　　　　　　　　　　14 tens

Write the 14.

$$\begin{array}{r} 2 \\ 3\,|7 \\ \times\ |4 \\ \hline 14\,|8 \end{array}$$

Note the pattern when the following numbers are multiplied.

Multiply the nonzero part of the factors.

Now attach the same number of zeros to the product as the total number of zeros in the factors.

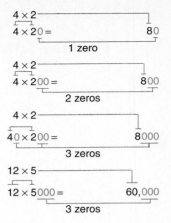

To multiply a number by 10, attach a zero to the right of the number.

$$376 \times 10 = 3760$$

To multiply a number by 100, attach two zeros to the right of the number.

$$59 \times 100 = 5900$$

To multiply a number by 1000, attach three zeros to the right of the number.

$$184 \times 1000 = 184{,}000$$

Multiply: 47×23

Step 1
Multiply by the ones' digit.

$$
\begin{array}{r}
47 \\
\times 23 \\
\hline
141
\end{array}
$$

Step 2
Multiply by the tens' digit.

$$
\begin{array}{r}
47 \\
\times 23 \\
\hline
141 \\
94
\end{array}
$$

Step 3
Add.

$$
\begin{array}{r}
47 \\
\times 23 \\
\hline
141 \\
94 \\
\hline
1081
\end{array}
$$

The last digit in the product is written in the ones' column.

The last digit in the product is written in the tens' column.

The place-value chart illustrates the placement of the products.

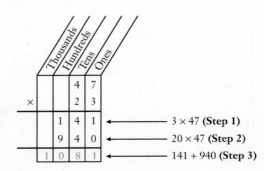

It is important to remember that the product of a number and zero is 0.

$$4 \times 0 = 0$$
$$0 \times 7 = 0$$
$$55 \times 0 = 0$$

Note how we place the products when multiplying by a factor that contains a zero.

Multiply: 439×206

$$
\begin{array}{r}
439 \\
\times 206 \\
\hline
2\,634 \\
0\,00 \quad 0 \times 439 \\
87\,8 \\
\hline
90{,}434
\end{array}
$$

When working the prob-
lem, we usually write
only one zero. Writing
this zero ensures the
proper placement of the
products.

$$
\begin{array}{r}
439 \\
\times 206 \\
\hline
2\,634 \\
87\,80 \\
\hline
90{,}434
\end{array}
$$

Repeated multiplication of the same factor can be written two ways.

$$3 \cdot 3 \cdot 3 \cdot 3 \cdot 3 \quad \text{or} \quad 3^5 \leftarrow \textbf{exponent}$$

The exponent, 5, indicates how many times the factor occurs in the multiplication. The 3 is called the **base**. The expression 3^5 is in **exponential notation**.

It is important to be able to read numbers that are written in exponential notation.

$6 = 6^1$ Read "six to the first power" or just "six." Usually the exponent 1 is not written.

$6 \cdot 6 = 6^2$ Read "six squared" or "six to the second power."

$6 \cdot 6 \cdot 6 = 6^3$ Read "six cubed" or "six to the third power."

$6 \cdot 6 \cdot 6 \cdot 6 = 6^4$ Read "six to the fourth power."

$6 \cdot 6 \cdot 6 \cdot 6 \cdot 6 = 6^5$ Read "six to the fifth power."

Each place value in the place-value chart can be expressed as 10 to a power.

Ten	=	10	=	10	$= 10^1$
Hundred	=	100	=	$10 \cdot 10$	$= 10^2$
Thousand	=	1000	=	$10 \cdot 10 \cdot 10$	$= 10^3$
Ten-thousand	=	10,000	=	$10 \cdot 10 \cdot 10 \cdot 10$	$= 10^4$
Hundred-thousand	=	100,000	=	$10 \cdot 10 \cdot 10 \cdot 10 \cdot 10$	$= 10^5$
Million		= 1,000,000	=	$10 \cdot 10 \cdot 10 \cdot 10 \cdot 10 \cdot 10$	$= 10^6$

To simplify a number expression containing an exponent, write the base as many times as indicated by the exponent; then multiply.

$$4^3 = 4 \cdot 4 \cdot 4 = 64$$

Example 1

Multiply: 829×603

Solution

$$
\begin{array}{r}
829 \\
\times\,603 \\
\hline
2\,487 \\
497\,40 \\
\hline
499{,}887
\end{array}
$$

You Try It 1

Multiply: 756×305

Your solution

Example 2

Simplify 5^3.

Solution

$5^3 = 5 \cdot 5 \cdot 5 = 125$

You Try It 2

Simplify 7^2.

Your solution

Solutions on p. A4

***O*bjective 1.3B**   *To divide whole numbers*

Division is used to separate objects into equal groups.

A grocer wants to distribute 24 new products equally on 4 shelves. From the diagram, we see that the grocer would place 6 products on each shelf.

The grocer's problem could be written:

$$
\underset{\substack{\textbf{Divisor}}}{\text{Number of shelves} \longrightarrow} \; \overset{\overset{\text{Number on each shelf}}{\overset{\textbf{Quotient}}{6}}}{4\overline{)24}} \longleftarrow \begin{array}{l}\text{Number of objects}\\ \textbf{Dividend}\end{array}
$$

Note that the quotient multiplied by the divisor equals the dividend.

$$\overset{\times \frown 6}{\underset{=}{4\overline{)24}}} \qquad \text{because} \qquad \boxed{\begin{array}{c}6\\ \text{Quotient}\end{array}} \times \boxed{\begin{array}{c}4\\ \text{Divisor}\end{array}} = \boxed{\begin{array}{c}24\\ \text{Dividend}\end{array}}$$

$$\overset{6}{9\overline{)54}} \qquad \text{because} \qquad 6 \quad \times \quad 9 \quad = \quad 54$$

Some important quotients follow.

Any whole number, except zero,
divided by itself is 1.

$$8\overline{)8} \qquad 14\overline{)14} \qquad 10\overline{)10}$$

Any whole number divided by
1 is the whole number.

$$1\overline{)9} \qquad 1\overline{)27} \qquad 1\overline{)10}$$

Zero divided by any other whole
number is zero.

$$7\overline{)0} \qquad 13\overline{)0} \qquad 10\overline{)0}$$

Division by zero is not allowed.

$$0\overline{)8}$$

There is no number
whose product
with 0 is 8.

When the dividend is a larger whole number, the digits in the quotient are found in
steps.

Divide $4\overline{)3192}$ and check.

Step 1

$$
\begin{array}{r}
7 \\
4\overline{)3192} \\
-28 \\
\hline
39
\end{array}
$$

Think $4\overline{)31}$.
Subtract 7×4.
Bring down the 9.

Step 2

$$
\begin{array}{r}
79 \\
4\overline{)3192} \\
-28 \\
\hline
39 \\
-36 \\
\hline
32
\end{array}
$$

Think $4\overline{)39}$.
Subtract 9×4.
Bring down the 2.

Step 3

$$
\begin{array}{r}
798 \\
4\overline{)3192} \\
-28 \\
\hline
39 \\
-36 \\
\hline
32 \\
-32 \\
\hline
0
\end{array}
$$

Think $4\overline{)32}$.
Subtract 8×4.

Step 4   Check:

$$
\begin{array}{r}
798 \\
\times \ 4 \\
\hline
3192
\end{array}
$$

The place-value chart
can be used to show why
this method works.

Sometimes it is not possible to separate objects into a whole number of equal groups.

A packer at a bakery has 14 muffins to pack into 3 boxes. Each box will hold 4 muffins. From the diagram, we see that after the packer places 4 muffins in each box, there are 2 muffins left over. The 2 is called the **remainder.**

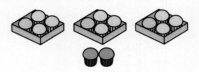

The packer's division problem could be written:

$$
\begin{array}{r}
4 \leftarrow \text{Number in each box (Quotient)} \\
\text{Number of boxes} \rightarrow 3\overline{)\ 14} \leftarrow \text{Total number of muffins (Dividend)} \\
\text{(Divisor)} \qquad \underline{-12} \\
2 \leftarrow \text{Number left over (Remainder)}
\end{array}
$$

The answer to a division problem with a remainder is frequently written:

$$3\overline{)14}^{\,4\ r2}$$

Note that

$$
\boxed{\begin{array}{c}4 \\ \text{Quotient}\end{array}} \times \boxed{\begin{array}{c}3 \\ \text{Divisor}\end{array}} + \boxed{\begin{array}{c}2 \\ \text{Remainder}\end{array}} = \boxed{\begin{array}{c}14 \\ \text{Dividend}\end{array}}
$$

When the divisor has more than one digit, estimate the quotient at each step by using the first digit of the divisor. If that guess does not work, lower the guess by 1 and try again.

Divide $34\overline{)1598}$ and check.

Step 1

$$
\begin{array}{r}
5 \\
34\overline{)\ 1598} \\
\underline{-170}
\end{array}
\qquad
\begin{array}{l}
5 \\
\text{Think } 3\overline{)15}. \\
\text{Subtract } 5 \times 34.
\end{array}
$$

Step 2

$$
\begin{array}{r}
4 \\
34\overline{)\ 1598} \\
\underline{-136} \\
238
\end{array}
\qquad
\text{Subtract } 4 \times 34.
$$

Because 170 is too large, lower the guess by 1 and try again.

Step 3

$$
\begin{array}{r}
47 \\
34\overline{)\ 1598} \\
\underline{-136} \\
238 \\
\underline{-238} \\
0
\end{array}
\qquad
\begin{array}{l}
7 \\
\text{Think } 3\overline{)23}. \\
\text{Subtract } 7 \times 34.
\end{array}
$$

Step 4 Check:

$$
\begin{array}{r}
47 \\
\times\ 34 \\
\hline
188 \\
141\ \\
\hline
1598
\end{array}
$$

Division can also be expressed by using the symbol ÷ or the symbol —, which are both read "divided by."

54 divided by 9 equals 6.

$$54 \div 9 = 6$$

$$\frac{54}{9} = 6$$

Example 3

Divide 7)2856 and check.

Solution

```
      408
7)  2856
   −28
     05      Think 7)5. Place 0 in quotient.
   − 0       Subtract 0 × 7.
     56      Bring down the 6.
   −56
      0
```

Check: 408 × 7 = 2856

Example 4

Divide 21,312 ÷ 56 and check.

Solution

```
       380 r32
56)  21,312       Think 5)21.
    −16 8          4 × 56 is too large.
     4 51          Try 3.
    −4 48
        32
      − 0
        32
```

Check: (380 × 56) + 32 =
      21,280   + 32 = 21,312

Example 5

Divide 386)206,149 and check.

Solution

```
        534 r25
386)  206,149
     −193 0
      13 14
     −11 58
       1 569
      −1 544
          25
```

Check:
(534 × 386) + 25 =
   206,124   + 25 = 206,149

You Try It 3

Divide 9)6345 and check.

Your solution

You Try It 4

Divide 18,359 ÷ 39 and check.

Your solution

You Try It 5

Divide 515)216,848 and check.

Your solution

Solutions on pp. A4–A5

*O*bjective 1.3C   *To find unit cost and total cost*

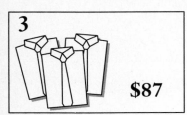

3

$87

Frequently stores advertise items for purchase in the following way:

> 3 shirts for $87   or   2 pounds of potatoes for 90¢

The **unit cost** is the cost of 1 shirt or 1 pound of potatoes. To find the unit cost, divide the total cost by the number of units.

3 shirts for $87	2 pounds of potatoes for 90¢
$87 \div 3 = 29$	$90 \div 2 = 45$
$29 is the cost of 1 shirt. The unit cost is $29 per shirt.	45¢ is the cost of 1 pound of potatoes. The unit cost is 45¢ per pound.

A coffee merchant wishes to purchase 50 pounds of java beans. Java beans are being sold for $5 per pound. What is the total cost of the merchant's purchase?

To find the **total cost,** multiply the unit cost by the number of units purchased.

Unit cost	×	Number of units	=	Total cost
5	×	50	=	250

The total cost is $250.

Example 6

A house painter wishes to purchase 24 gallons of exterior latex paint. The paint costs $12 per gallon. Find the total cost of the 24 gallons of paint.

Strategy

To find the total cost, multiply the unit cost ($12) by the number of units (24).

Solution

$$\begin{array}{r} 12 \\ \times\,24 \\ \hline 48 \\ 24 \\ \hline 288 \end{array}$$

The total cost is $288.

You Try It 6

The owner of a nursery wishes to purchase 35 pine saplings. Pine saplings cost $6 each. Find the total cost of the 35 pine saplings.

Your strategy

Your solution

Solution on p. A5

*E*XERCISE 1.3A

Multiply.

1.   89
    × 7

2.   45
    × 9

3.   623
    × 4

4.   802
    × 5

5.   607
    × 9

6.   4780
    ×   4

7. 48,253
    ×   7

8.   27
    ×72

9.   95
    ×33

10. 727
    ×60

11. 588
    ×75

12. 8279
    × 46

13. 6938
    × 78

14. 3009
    × 35

15. 6003
    × 57

16. 3987
    × 29

17. 4765
    × 37

18. 607
    ×406

19. 809
    ×503

20. 312
    ×134

21. 423
    ×427

22. 386
    ×759

23. 2675
    × 487

24. 3985
    × 364

1. _____
2. _____
3. _____
4. _____
5. _____
6. _____
7. _____
8. _____
9. _____
10. _____
11. _____
12. _____
13. _____
14. _____
15. _____
16. _____
17. _____
18. _____
19. _____
20. _____
21. _____
22. _____
23. _____
24. _____

Simplify.

25. 2^3	**26.** 5^2	**27.** 9^2	**28.** 4^3

29. 0^4	**30.** 1^6	**31.** 3^4	**32.** 6^3

33. 10^2	**34.** 10^4	**35.** 10^3	**36.** 10^5

EXERCISE 1.3B

Divide.

37. $6)\overline{19,254}$	**38.** $4)\overline{39,200}$	**39.** $7)\overline{632}$

40. $4)\overline{363}$	**41.** $8)\overline{1635}$	**42.** $7)\overline{9432}$

43. $9)\overline{7004}$	**44.** $7)\overline{6001}$	**45.** $4)\overline{15,300}$

46. $7)\overline{43,500}$	**47.** $27)\overline{96}$	**48.** $44)\overline{82}$

25. _____

26. _____

27. _____

28. _____

29. _____

30. _____

31. _____

32. _____

33. _____

34. _____

35. _____

36. _____

37. _____

38. _____

39. _____

40. _____

41. _____

42. _____

43. _____

44. _____

45. _____

46. _____

47. _____

48. _____

49. 41)897   **50.** 32)693   **51.** 44)8821

52. 19)3859   **53.** 22)98,644   **54.** 77)83,629

55. 206)3097   **56.** 87)4911   **57.** 169)8542

58. 456)7723   **59.** 223)8927   **60.** 467)9344

EXERCISE 1.3C

Solve.

61. The cost for 4 sponges is 88¢. Find the cost of 1 sponge.

62. Two boxes of light bulbs cost 98¢. Find the cost of 1 box of light bulbs.

63. Two bottles of dry toner for a copy machine cost $34. Find the unit cost.

64. A package of 6 printer ribbons costs $84. Find the unit cost.

65. The cost for an order of 8 desks is $1160. Find the unit cost.

66. The cost for an order of 5 computer terminals is $8250. Find the unit cost.

67. The total cost of manufacturing 775 aluminum baseball bats is $9300. Find the unit cost.

68. The total cost of manufacturing 825 canvas tents is $12,375. Find the unit cost.

49. _____
50. _____
51. _____
52. _____
53. _____
54. _____
55. _____
56. _____
57. _____
58. _____
59. _____
60. _____
61. _____
62. _____
63. _____
64. _____
65. _____
66. _____
67. _____
68. _____

69. Clear redwood lumber costs $8 per foot. How much would a construction company pay for 275 feet of clear redwood lumber?

70. An installer of floor tiles finds that the unit cost of the floor tiles requested by a customer is $6. How much would the installer pay for 350 floor tiles?

71. A company orders 150 boxes of memo paper. Each box costs $12. Find the total cost of the order.

72. The King Company orders 250 boxes of letterhead stationery. Each box costs $15. Find the total cost of the order.

73. A contractor quotes the cost of work on a new house, which is to have 2100 square feet of floor space, at $85 per square foot. Find the total cost of the contractor's work on the house.

74. The Acton Corporation is considering purchasing a 35-acre tract of land on which to build new facilities. The price is $2575 per acre. Find the total cost of the land.

75. A package of 10 double-sided, $3\frac{1}{2}$-inch computer disks costs $40. Find the total cost when 50 disks are ordered.

76. Replacement printer ribbons are sold in packages of 3 for $16. Find the total cost for 15 ribbons.

77. The cost of tomatoes is 3 pounds for 99¢. Find the cost of 2 pounds of tomatoes.

78. The cost of one brand of tires is 2 for $240. Find the cost of 3 tires.

79. The owner of a motel placed an order for 35 dressers with a furniture manufacturer. The cost for each of the first 25 dressers ordered was $135. Each additional dresser ordered cost only $115. Find the total cost of the 35 dressers.

80. The owner of a music store ordered 65 electronic keyboards from a manufacturer of musical instruments. The cost for each of the first 50 keyboards ordered was $185. Each additional keyboard cost only $145. Find the total cost of the 65 electronic keyboards.

69. _____

70. _____

71. _____

72. _____

73. _____

74. _____

75. _____

76. _____

77. _____

78. _____

79. _____

80. _____

CALCULATORS

Estimation An important skill in business is the ability to determine whether a calculation is reasonable. One method of determining whether a calculation is reasonable is to use estimation. An **estimate** is an approximation.

Estimation is especially valuable when using a calculator. Suppose you are adding 1497 and 2568 on a calculator. You enter the number 1497 correctly but inadvertently enter 256 instead of 2568 for the second addend. The sum reads 1753. If you quickly make an estimate of the answer, you can determine that the sum 1753 is not reasonable and that an error has been made.

$$\begin{array}{r} 1497 \\ +\ 2568 \\ \hline 4065 \end{array} \qquad \begin{array}{r} 1497 \\ +\ 256 \\ \hline 1753 \end{array}$$

To estimate the answer to a calculation, round each number to the highest place value of the number; that is, the first digit of each number will be nonzero and all other digits will be zero. Perform the calculation using the rounded numbers.

$$\begin{array}{rcl} 1497 & \longrightarrow & 1000 \\ 2568 & \longrightarrow & +\ 3000 \\ & & \hline 4000 \end{array}$$

As shown above, the sum 4000 is an estimate of the sum of 1497 and 2568; it is very close to the actual sum 4065. 4000 is not close to the incorrectly calculated sum 1753.

Estimate the difference between 27,843 and 19,206.

Round each number to the nearest ten-thousand.
Subtract the rounded numbers.

$$\begin{array}{rcl} 27{,}843 & \longrightarrow & 30{,}000 \\ 19{,}206 & \longrightarrow & -\ 20{,}000 \\ & & \hline 10{,}000 \end{array}$$

10,000 is an estimate of the difference between 27,843 and 19,206.

Estimate the product of 387 and 214.

Round each number to the nearest hundred.

Multiply the rounded numbers.

$$\begin{array}{rcl} 387 & \longrightarrow & 400 \\ 214 & \longrightarrow & 200 \end{array}$$

$$400 \cdot 200 = 80{,}000$$

80,000 is an estimate of the product of 387 and 214.

Estimate by rounding. Then use a calculator to find the exact answer.

1. $6742 + 8298$

2. $5426 + 1732$

3. $7355 - 5219$

4. $8953 - 2217$

5. 3467×359

6. $8745 \cdot 63$

1. _____

2. _____

3. _____

4. _____

5. _____

6. _____

Estimate by rounding. Then use a calculator to find the exact answer.

7. $36,484 \div 47$

8. $62,176 \div 58$

9. $792,085 + 416,832$

10. $23,774 + 38,026$

11. $59,126 - 20,843$

12. $63,051 - 29,478$

13. $(39,246)(29)$

14. $64,409 \cdot 67$

15. $389,804 \div 76$

16. $637,051 \div 29$

17. 8941×726

18. $2837(216)$

19. $332,174 \div 219$

20. $632,034 \div 219$

21. $224,196 - 98,531$

22. $873,925 - 28,744$

23. $387 + 295 + 614$

24. $528 + 163 + 974$

Solve.

25. A buyer for a department store purchased 215 suits at $83 each. Estimate the total cost of the order.

26. A sales executive earned commissions of $4875, $3925, and $5165 during a three-month period. Estimate the total of the executive's commissions for the three-month period.

27. A car company produced 3285 cars in April, 2714 cars in May, and 2182 cars in June. Estimate the total number of cars manufactured by the company during the three months.

28. The odometer on your car read 58,376 at this time last year. It now reads 77,912. Estimate the number of miles your car has been driven during the past year.

29. A limited partnership consisting of 285 investors sold a piece of land for $5,957,250. Estimate the amount received by each investor.

30. There are 52 weeks in a year. Is this an exact figure or an approximation?

7. _____

8. _____

9. _____

10. _____

11. _____

12. _____

13. _____

14. _____

15. _____

16. _____

17. _____

18. _____

19. _____

20. _____

21. _____

22. _____

23. _____

24. _____

25. _____

26. _____

27. _____

28. _____

29. _____

30. _____

BUSINESS CASE STUDY

Entrepreneur

Entrepreneurs are people who risk their time, effort, and money to start and operate a business. Suppose you have decided to become an entrepreneur and open a bicycle repair shop. The cost to rent the space you are interested in is $700 per month. You have been told that utilities (heat, water, and electricity) will cost approximately $115 per month. All maintenance work, such as painting the exterior of the building, will be provided by the owner of the building. The installation fee for the telephone will be $55. The monthly telephone bill will be about $30. Insurance coverage for yourself and for your business will cost $1400 a year.

Because you have been working on your own bikes for several years, you have most of the tools required for the business. You plan to spend only $400 to purchase some additional tools necessary to repair other models of bicycles. You have estimated that you will need to spend $150 per month to purchase parts that are necessary to repair bicycles.

You are convinced that advertising your business is essential. You have decided that during the first week of operation, you will place a half-page ad in your local newspaper, which is published weekly. The cost for a half-page ad is $150. Each week after the first, a 5-inch ad will appear on the page listing business services. The weekly cost is $4 per inch.

For at least the first year of operation, you plan on running the business by yourself and therefore will not be hiring any employees. This will save the expense of wages and benefits, but it will require that you take time from repair jobs in order to answer the telephone, greet customers, order supplies, and so on. Having taken business courses in college, you feel confident that you can prepare your own tax forms, thereby saving the expense of paying an accountant to do them for you.

You plan to charge customers for both parts and labor. The amount charged a customer for a part used to repair a bike will be the same price that you paid to purchase it. You will charge a customer $25 per hour for your labor. Your policy at the shop will be that a customer must pay for the repair work at the time the bicycle is picked up. Therefore, you are not concerned about having unpaid bills.

1. **Profit** is what remains after all business expenses have been subtracted from the amount of money received as a result of doing business. If expenses are greater than the amount received, the business experiences a **loss**. Assuming that you devote 25 hours each week to repairing bicycles, what would be your profit or loss during the first year of operation? What is the result if you don't have enough customers to keep you busy 25 hours each week, and your repair work requires only 15 hours each week?

2. Assume you are presently employed and earning $22,500 per year. How does your possible profit compare with this figure?

3. What are some of the disadvantages of going into the bicycle repair business? For example, consider vacation time, medical coverage, time devoted to work, risk, and seasonal changes in the demand for your services.

4. What are some of the advantages of going into business for yourself?

CHAPTER SUMMARY

Key Words   The **whole numbers** are 0, 1, 2, 3, 4, 5, 6, 7, 8, 9, 10, 11, 12, 13, and so on. (Objective 1.1A)

When a whole number is written using the digits 0, 1, 2, 3, 4, 5, 6, 7, 8, and 9, it is said to be in **standard form.** The position of each digit in the number determines the digit's **place value.** (Objective 1.1A)

Giving an approximate value of an exact answer is called **rounding.** (Objective 1.1B)

Addition is the process of finding the total of two or more numbers. The two numbers being added are called **addends.** The answer is the **sum.** (Objective 1.2A)

Subtraction is the process of finding the difference between two numbers. The **subtrahend** is subtracted from the **minuend.** The answer is the **difference.** (Objective 1.2B)

Multiplication is repeated addition of the same number. The numbers that are multiplied are called **factors.** The answer is the **product.** (Objective 1.3A)

The expression 5^3 is in **exponential notation.** The **exponent,** 3, indicates how many times the base, 5, occurs in the multiplication. (Objective 1.3A)

Division is used to separate objects into equal groups. The **dividend** is divided by the **divisor.** The answer is the **quotient.** The **remainder** is the number left over when the divisor does not divide into the dividend evenly. (Objective 1.3B)

A **quota** establishes either an upper limit or a lower limit. A sales quota establishes a minimum amount of sales expected during a given period. (Objective 1.2D)

Unit cost is the cost for one item. (Objective 1.3C)

Essential Rules   **To round a number to a given place value:** If the digit to the right of the given place value is less than 5, that digit and all digits to the right are replaced by zeros. If the digit to the right of the given place value is greater than or equal to 5, increase the digit in the given place value by 1, and replace all other digits to the right by zeros. (Objective 1.1B)

Check for a subtraction problem: subtrahend + difference = minuend (Objective 1.2B)

To multiply by a power of 10, multiply the nonzero part of the factors. Then attach the same number of zeros to the product as the total number of zeros in the factors. (Objective 1.3A)

The product of a number and zero is zero. (Objective 1.3A)

Important quotients: Any whole number, except zero, divided by itself is 1. Any whole number divided by 1 is the whole number. Zero divided by any other whole number is zero. Division by zero is not allowed. (Objective 1.3B)

Check for a division problem: (quotient $\times$ divisor) + remainder = dividend (Objective 1.3B)

Total cost is found by multiplying the unit cost by the number of units purchased:

Unit cost $\times$ Number of units = Total cost (Objective 1.3C)

1. Write 207,068 in words.

2. Write one million two hundred four thousand six in standard form.

3. Round 74,965 to the nearest hundred.

4. Add: 25,492
   + 71,306

5. Add: 89,756
      9,094
   + 37,065

6. Add: 87,256 + 3095 + 9981

7. Subtract: 17,495
      − 8,162

8. Subtract: 20,736
      − 9,854

9. Subtract: 9504 − 7819

10. Multiply: 90,763
      ×   8

11. Multiply: 9736
      × 704

12. Multiply: 1400 × 20

1. _____

_____

_____

2. _____

3. _____

4. _____

5. _____

6. _____

7. _____

8. _____

9. _____

10. _____

11. _____

12. _____

13. Simplify: 8^2

14. Divide: $5624 \div 8$

15. Divide: $7\overline{)60,972}$

16. Divide: $97\overline{)108,764}$

13. _____

14. _____

15. _____

16. _____

17. _____

18. _____

19. _____

20. _____

17. A manufacturer of sports equipment had an inventory of 1890 skateboards before filling an order for 245 skateboards. Find the manufacturer's current inventory of skateboards.

18. A sales representative has a sales quota of 2500 units for the first three months of the year. The representative sold 857 units in January and 796 units in February. How many units must the representative sell in March in order to meet the quota?

19. The total cost of manufacturing 750 briefcases was $12,750. Find the unit cost.

20. A company orders 175 boxes of letterhead stationery at a cost of $13 per box. Find the total cost of the order.

2

Fractions and Business Applications

*O*BJECTIVES

2.1A   To write an improper fraction as a mixed number or a whole number, and a mixed number as an improper fraction

2.1B   To build equivalent fractions

2.1C   To reduce fractions

2.2A   To find the least common multiple (LCM)

2.2B   To add proper fractions

2.2C   To add whole numbers, mixed numbers, and fractions

2.2D   To subtract proper fractions

2.2E   To subtract whole numbers, mixed numbers, and fractions

2.2F   To solve problems involving shares of stock

2.3A   To multiply proper fractions

2.3B   To multiply whole numbers, mixed numbers, and fractions

2.3C   To divide proper fractions

2.3D   To divide whole numbers, mixed numbers, and fractions

2.3E   To solve problems involving hourly wage

35

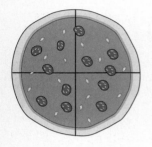

"**W**HY DO I need to learn this?" Students ask this question most frequently when they find the material difficult.

Many students consider operations with fractions difficult, especially adding and subtracting fractions with different denominators. The difficulty often lies in the fact that in any one problem, there may be a number of steps to be performed, and the purpose of each step in the process must be kept in mind as the problem is worked.

Sometimes students confuse the process of adding fractions and the process of multiplying fractions. In order to be added together, fractions must have the same denominator. Two fractions need not have the same denominator to be multiplied. When you find yourself uncertain, use a concrete example to help you determine how to proceed. For example, suppose you and a friend share a pizza, and you eat 1/4 of the pizza and your friend eats 1/4 of the pizza. To determine how much of the pizza was eaten by you both, you must add the two fractions. Would it be reasonable to multiply the numerators and multiply the denominators? The answer would be 1/16, which is not a reasonable answer. The correct answer is calculated by adding the numerators and placing the sum over the common denominator, 4; the two of you ate 2/4, or 1/2, of the pizza.

Or suppose you split a pizza with a friend; your share is 1/2 of the pizza. If you eat 1/2 of your share, how much of the whole pizza have you eaten? You must multiply the fractions. Would it be reasonable to add the numerators and place the sum over the common denominator, 2? The answer would be 1, the whole pizza, which is not a reasonable answer. The correct answer is calculated by multiplying the numerators and multiplying the denominators; you ate 1/4 of the whole pizza.

In this chapter, you will be applying the concepts of fractions to solving problems that involve shares of stock and problems that involve overtime pay. You will be working with fractions when you study concepts presented in later chapters of this text, such as

Converting fractions and decimals
Converting fractions and percents
Allocation and proportion
Formulas
Interest rates

Therefore, the effort you invest in this chapter will be rewarded as you continue your work in this course.

SECTION 2.1   Introduction to Fractions

Objective 2.1A *To write an improper fraction as a mixed number or a whole number, and a mixed number as an improper fraction*

A **fraction** can represent the number of equal parts of a whole.

The shaded portion of the circle is represented by the fraction $\frac{4}{7}$. Four sevenths of the circle are shaded.

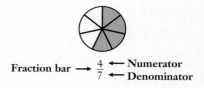

Fraction bar → $\frac{4}{7}$ ← **Numerator** / ← **Denominator**

Each part of a fraction has a name. The **fraction bar** separates the **numerator** (the number above the fraction bar) from the **denominator** (the number below the fraction bar).

A **proper fraction** is a fraction less than 1. The numerator of a proper fraction is smaller than the denominator. The shaded portion of the circle can be represented by the proper fraction $\frac{3}{4}$.

$\frac{3}{4}$

A **mixed number** is a number greater than 1 that consists of a whole-number part and a fractional part. The shaded portion of the circles can be represented by the mixed number $2\frac{1}{4}$.

$2\frac{1}{4}$

An **improper fraction** is a fraction greater than or equal to 1. The numerator of an improper fraction is greater than or equal to the denominator. The shaded portion of the circles can be represented by the improper fraction $\frac{9}{4}$. The shaded portion of the square can be represented by the improper fraction $\frac{4}{4}$.

$\frac{9}{4}$

$\frac{4}{4}$

Note from the diagram that the mixed number $2\frac{3}{5}$ and the improper fraction $\frac{13}{5}$ represent the shaded portion of the circles.

$$2\frac{3}{5} = \frac{13}{5}$$

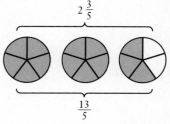

$2\frac{3}{5}$

$\frac{13}{5}$

An improper fraction can be written as a mixed number.

Write $\frac{13}{5}$ as a mixed number.

Step 1	**Step 2**	**Step 3**
Divide the numerator by the denominator.	To write the fractional part of the mixed number, write the remainder over the divisor.	Write the answer.

$$5\overline{)\begin{array}{c}2\\13\end{array}}$$
$$\underline{-10}$$
$$3$$

$$\begin{array}{c}2\frac{3}{5}\\5\overline{)13}\end{array}$$
$$\underline{-10}$$
$$3$$

$$\frac{13}{5} = 2\frac{3}{5}$$

To write a mixed number as an improper fraction, multiply the denominator of the fractional part by the whole-number part. The sum of this product and the numerator of the fractional part is the numerator of the improper fraction. The denominator remains the same.

Write $7\frac{3}{8}$ as an improper fraction.

$$7\frac{3}{8} = \frac{(8 \times 7) + 3}{8} = \frac{56 + 3}{8} = \frac{59}{8} \qquad 7\frac{3}{8} = \frac{59}{8}$$

Example 1
Write $\frac{21}{4}$ as a mixed number.

Solution

$$4\overline{)\begin{array}{c}5\\21\end{array}}$$
$$\underline{-20}$$
$$1$$

$$\frac{21}{4} = 5\frac{1}{4}$$

Example 2
Write $\frac{18}{6}$ as a whole number.

Solution

$$6\overline{)\begin{array}{c}3\\18\end{array}}$$
$$\underline{-18}$$
$$0$$

Note: The remainder is zero.

$$\frac{18}{6} = 3$$

Example 3
Write $21\frac{3}{4}$ as an improper fraction.

Solution

$$21\frac{3}{4} = \frac{84 + 3}{4} = \frac{87}{4}$$

You Try It 1
Write $\frac{22}{5}$ as a mixed number.

Your solution

You Try It 2
Write $\frac{28}{7}$ as a whole number.

Your solution

You Try It 3
Write $14\frac{5}{8}$ as an improper fraction.

Your solution

Solutions on pp. A5–A6

Objective 2.1B   *To build equivalent fractions*

Equal fractions with different denominators are called **equivalent fractions.**

$\frac{4}{6}$ is equivalent to $\frac{2}{3}$.

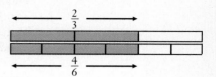

To *build* a fraction to an equivalent fraction, multiply the numerator and denominator by the same number.

$$\frac{2}{3} = \frac{2 \cdot 2}{3 \cdot 2} = \frac{4}{6}$$

$\frac{2}{3}$ was built to the equivalent fraction $\frac{4}{6}$.

Build a fraction that is equivalent to $\frac{5}{8}$ and has a denominator of 32.

$$\frac{5}{8} = \frac{?}{32}$$

Step 1   Divide the larger denominator by the smaller. $32 \div 8 = 4$

Step 2   Multiply the numerator and denominator of the given fraction by the quotient (4).

$$\frac{5 \cdot 4}{8 \cdot 4} = \frac{20}{32}$$

$\frac{20}{32}$ is equivalent to $\frac{5}{8}$.

Example 4

Build a fraction that is equivalent to $\frac{2}{3}$ and has a denominator of 42.

Solution

$$\frac{2}{3} = \frac{?}{42}$$

$42 \div 3 = 14$ $\dfrac{2 \cdot 14}{3 \cdot 14} = \dfrac{28}{42}$

$\frac{28}{42}$ is equivalent to $\frac{2}{3}$.

You Try It 4

Build a fraction that is equivalent to $\frac{3}{5}$ and has a denominator of 45.

Your solution

Example 5

Build a fraction that is equivalent to 4 and has a denominator of 12.

Solution

Write 4 as $\frac{4}{1}$. $\dfrac{4}{1} = \dfrac{?}{12}$

$12 \div 1 = 12$ $\dfrac{4 \cdot 12}{1 \cdot 12} = \dfrac{48}{12}$

$\frac{48}{12}$ is equivalent to 4.

You Try It 5

Build a fraction that is equivalent to 6 and has a denominator of 18.

Your solution

Solutions on p. A6

Objective 2.1C   *To reduce fractions*

Whole-number factors of a number divide that number evenly (there is no remainder).

Because they divide 6 evenly, 1, 2, 3, and 6 are whole-number factors of 6.

$$1\overline{)6}^{\,6} \qquad 2\overline{)6}^{\,3} \qquad 3\overline{)6}^{\,2} \qquad 6\overline{)6}^{\,1}$$

A number that is a factor of two or more numbers is a **common factor** of those numbers. Because 3 divides both 6 and 9 evenly, 3 is a common factor of 6 and 9.

A fraction is in simplest form when the only common factor of the numerator and the denominator is 1. The fractions $\frac{3}{8}$, $\frac{4}{9}$, and $\frac{5}{11}$ are in simplest form.

To *reduce* a fraction to **simplest form,** divide the numerator and denominator by a common factor so that their only common factor is 1.

$$\frac{4}{6} = \frac{4 \div 2}{6 \div 2} = \frac{2}{3}$$

$$\frac{18}{30} = \frac{18 \div 6}{30 \div 6} = \frac{3}{5}$$

An improper fraction should be changed to a mixed number.

$$\frac{22}{6} = \frac{22 \div 2}{6 \div 2} = \frac{11}{3} = 3\frac{2}{3}$$

The following rules are helpful in finding factors of a number.

A number can be divided by 2 if the ones' digit is 0, 2, 4, 6, or 8.
A number can be divided by 3 if the sum of the digits is divisible by 3.
A number can be divided by 4 if the last two digits are divisible by 4.
A number can be divided by 5 if the ones' digit is 0 or 5.
A number can be divided by 9 if the sum of the digits is divisible by 9.
A number can be divided by 10 if the ones' digit is 0.

For example, 357 is divisible by 3 because $3 + 5 + 7 = 15$ and 15 is divisible by 3. 928 is divisible by 4 because 28 is divisible by 4.

Example 6

Reduce $\frac{15}{40}$ to simplest form.

Solution

$$\frac{15}{40} = \frac{15 \div 5}{40 \div 5} = \frac{3}{8}$$

You Try It 6

Reduce $\frac{16}{24}$ to simplest form.

Your solution

Example 7

Reduce $\frac{30}{12}$ to simplest form.

Solution

$$\frac{30}{12} = \frac{30 \div 6}{12 \div 6} = \frac{5}{2} = 2\frac{1}{2}$$

You Try It 7

Reduce $\frac{48}{36}$ to simplest form.

Your solution

Solutions on p. A6

EXERCISE 2.1A

Express the shaded portion of the circles as an improper fraction and as a mixed number.

1.

2.

Write the improper fraction as a mixed number or a whole number.

3. $\dfrac{11}{4}$

4. $\dfrac{7}{3}$

5. $\dfrac{20}{4}$

6. $\dfrac{9}{8}$

7. $\dfrac{23}{10}$

8. $\dfrac{48}{16}$

9. $\dfrac{16}{3}$

10. $\dfrac{29}{2}$

11. $\dfrac{16}{1}$

12. $\dfrac{17}{8}$

13. $\dfrac{31}{16}$

14. $\dfrac{12}{5}$

Write the mixed number as an improper fraction.

15. $2\dfrac{1}{3}$

16. $4\dfrac{2}{3}$

17. $6\dfrac{1}{2}$

18. $6\dfrac{5}{6}$

19. $7\dfrac{3}{8}$

20. $9\dfrac{1}{4}$

21. $10\dfrac{1}{2}$

22. $15\dfrac{1}{8}$

23. $3\dfrac{7}{9}$

24. $2\dfrac{5}{8}$

25. $1\dfrac{3}{8}$

26. $4\dfrac{5}{9}$

1. _____
2. _____
3. _____
4. _____
5. _____
6. _____
7. _____
8. _____
9. _____
10. _____
11. _____
12. _____
13. _____
14. _____
15. _____
16. _____
17. _____
18. _____
19. _____
20. _____
21. _____
22. _____
23. _____
24. _____
25. _____
26. _____

EXERCISE 2.1B

Build an equivalent fraction with the given denominator.

27. $\dfrac{1}{4} = \dfrac{}{16}$

28. $\dfrac{5}{9} = \dfrac{}{81}$

29. $\dfrac{7}{11} = \dfrac{}{33}$

30. $\dfrac{5}{8} = \dfrac{}{32}$

31. $5 = \dfrac{}{25}$

32. $9 = \dfrac{}{4}$

33. $\dfrac{11}{15} = \dfrac{}{60}$

34. $\dfrac{5}{9} = \dfrac{}{36}$

35. $\dfrac{7}{9} = \dfrac{}{45}$

36. $\dfrac{11}{12} = \dfrac{}{36}$

37. $\dfrac{15}{16} = \dfrac{}{64}$

38. $\dfrac{11}{18} = \dfrac{}{54}$

EXERCISE 2.1C

Reduce the fraction to simplest form.

39. $\dfrac{2}{14}$

40. $\dfrac{50}{75}$

41. $\dfrac{40}{36}$

42. $\dfrac{75}{25}$

43. $\dfrac{8}{60}$

44. $\dfrac{12}{35}$

45. $\dfrac{24}{40}$

46. $\dfrac{12}{16}$

47. $\dfrac{4}{32}$

48. $\dfrac{60}{100}$

49. $\dfrac{36}{16}$

50. $\dfrac{80}{45}$

27. _____

28. _____

29. _____

30. _____

31. _____

32. _____

33. _____

34. _____

35. _____

36. _____

37. _____

38. _____

39. _____

40. _____

41. _____

42. _____

43. _____

44. _____

45. _____

46. _____

47. _____

48. _____

49. _____

50. _____

SECTION 2.2   Addition and Subtraction of Fractions and Mixed Numbers

Objective 2.2A   *To find the least common multiple (LCM)*

The **multiples** of a number are the products of that number and the numbers 1, 2, 3, 4, 5, and so on.

$3 \times 1 = 3$
$3 \times 2 = 6$
$3 \times 3 = 9$ Some of the multiples of 3 are 3, 6, 9, 12, and 15.
$3 \times 4 = 12$
$3 \times 5 = 15$

A number that is a multiple of two or more numbers is a **common multiple** of those numbers.

The multiples of 4 are 4, 8, 12, 16, 20, 24, 28, 32, 36, ...
The multiples of 6 are 6, 12, 18, 24, 30, 36, 42, ...
Some common multiples of 4 and 6 are 12, 24, and 36.

The **least common multiple** (LCM) is the smallest common multiple of two or more numbers. The least common multiple of 4 and 6 is 12.

Listing the multiples of each number is one way to find the LCM. Another way to find the LCM uses prime numbers. A number is **prime** if its only whole-number factors are 1 and itself. The number 7 is prime because its only factors are 1 and 7; 6 is not a prime number because it has factors of 2 and 3. The number 1 is not considered a prime number. Therefore, it is not included in the following list of prime numbers less than 25.

2, 3, 5, 7, 11, 13, 17, 19, 23

To find the LCM of 4 and 6, list the numbers in a row.

$$4 \quad 6$$

Divide by a prime number that divides either of the numbers.

$$2 \,\lfloor\, 4 \quad 6$$
$$\overline{2 \quad 3}$$

Here 2 divides both 4 and 6.

Continue dividing by prime numbers until the last row is all 1's. If a prime number does not divide one of the numbers evenly, repeat the number on the next row.

$$2 \,\lfloor\, 4 \quad 6$$
$$2 \,\lfloor\, 2 \quad 3$$
$$1 \quad 3$$

Here 2 divides 2 but not 3. Repeat the 3 in this row.

$$2 \,\lfloor\, 4 \quad 6$$
$$2 \,\lfloor\, 2 \quad 3$$
$$3 \,\lfloor\, 1 \quad 3$$
$$1 \quad 1$$

Here 3 divides 3 but not 1. Repeat the 1 in this row. This last row is all 1's.

The LCM is the product of the numbers in the left-hand column.

$2 \cdot 2 \cdot 3 = 12$
The LCM of 4 and 6 is 12.

Example 1

Find the LCM of 3, 4, and 12.

Solution

$$
\begin{array}{c|ccc}
3 & 3 & 4 & 12 \\
\hline
2 & 1 & 4 & 4 \\
\hline
2 & 1 & 2 & 2 \\
\hline
& 1 & 1 & 1
\end{array}
$$
The LCM $= 3 \cdot 2 \cdot 2 = 12$.

You Try It 1

Find the LCM of 2, 6, and 8.

Your solution

Solution on p. A6

Objective 2.2B   *To add proper fractions*

Fractions with the same denominator are added by adding the numerators and placing the sum over the common denominator. After adding, reduce the fraction to simplest form.

Add: $\dfrac{2}{7} + \dfrac{4}{7}$

$$
\begin{array}{r}
\dfrac{2}{7} \\
+\dfrac{4}{7} \\
\hline
\dfrac{6}{7}
\end{array}
$$
Add the numerators and place the sum over the common denominator.

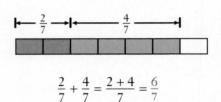

$$\frac{2}{7} + \frac{4}{7} = \frac{2+4}{7} = \frac{6}{7}$$

To add fractions with different denominators, first rewrite the fractions as equivalent fractions with a common denominator. The common denominator is the least common multiple (LCM) of the numbers in the denominators and is called the **lowest common denominator** (LCD).

Add: $\dfrac{1}{2} + \dfrac{1}{3}$

Step 1
Find the LCD of the denominators 2 and 3.

LCD = 6

Step 2
Build equivalent fractions using the LCD.

$$\frac{1}{2} = \frac{3}{6}$$

$$+\frac{1}{3} = \frac{2}{6}$$

Step 3
Add the fractions.

$$
\begin{array}{r}
\dfrac{1}{2} = \dfrac{3}{6} \\
+\dfrac{1}{3} = \dfrac{2}{6} \\
\hline
\dfrac{5}{6}
\end{array}
$$

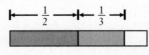

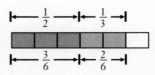

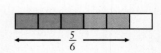

Example 2

Add: $\dfrac{5}{8} + \dfrac{2}{3}$

Solution

$\dfrac{5}{8} = \dfrac{15}{24}$ The LCD of 8 and 3 is 24.

$+\dfrac{2}{3} = \dfrac{16}{24}$

$\quad\quad \dfrac{31}{24} = 1\dfrac{7}{24}$

Example 3

Add: $\dfrac{2}{3} + \dfrac{3}{5} + \dfrac{5}{6}$

Solution

$\dfrac{2}{3} = \dfrac{20}{30}$ The LCD of 3, 5, and 6 is 30.

$\dfrac{3}{5} = \dfrac{18}{30}$

$+\dfrac{5}{6} = \dfrac{25}{30}$

$\quad\quad \dfrac{63}{30} = 2\dfrac{3}{30} = 2\dfrac{1}{10}$

Add: $\dfrac{7}{8} + \dfrac{3}{5}$

Your solution

Add: $\dfrac{3}{4} + \dfrac{4}{5} + \dfrac{5}{8}$

Your solution

Solutions on p. A6

Objective 2.2C   *To add whole numbers, mixed numbers, and fractions*

The sum of a whole number and a fraction is a mixed number.

Add: $2 + \dfrac{2}{3}$

$\boxed{2} + \dfrac{2}{3} = \boxed{\dfrac{6}{3}} + \dfrac{2}{3} = \dfrac{8}{3} = 2\dfrac{2}{3}$

To add a whole number and a mixed number, write the fraction and then add the whole numbers.

Add: $7\dfrac{2}{5} + 4$

Step 1   Write the fraction.

$\begin{array}{r} 7\dfrac{2}{5} \\ +\,4 \\ \hline \dfrac{2}{5} \end{array}$

Step 2   Add the whole numbers.

$\begin{array}{r} 7\dfrac{2}{5} \\ +\,4 \\ \hline 11\dfrac{2}{5} \end{array}$

To add two mixed numbers, add the fractional parts and then add the whole numbers. Remember to reduce the sum to simplest form.

Add: $5\dfrac{4}{9} + 6\dfrac{14}{15}$

Step 1   Add the fractional parts.

$$5\dfrac{4}{9} = 5\dfrac{20}{45}$$ The LCD of 9 and
$$+\,6\dfrac{14}{15} = 6\dfrac{42}{45}$$ 15 is 45.
$$\overline{\qquad\qquad\dfrac{62}{45}}$$

Step 2   Add the whole numbers.

$$5\dfrac{4}{9} = 5\dfrac{20}{45}$$
$$+\,6\dfrac{14}{15} = 6\dfrac{42}{45}$$
$$\overline{\qquad 11\dfrac{62}{45} = 11 + 1\dfrac{17}{45} = 12\dfrac{17}{45}}$$

Example 4

Add: $17 + 3\dfrac{3}{8}$

Solution

$$\begin{aligned}17\\ +\ 3\dfrac{3}{8}\\ \overline{20\dfrac{3}{8}}\end{aligned}$$

You Try It 4

Add: $29 + 17\dfrac{5}{12}$

Your solution

Example 5

Add: $5\dfrac{2}{3} + 11\dfrac{5}{6} + 12\dfrac{7}{9}$

Solution

$$5\dfrac{2}{3} = \ 5\dfrac{12}{18}$$ The LCD of 3, 6, and 9 is 18.
$$11\dfrac{5}{6} = 11\dfrac{15}{18}$$
$$+\,12\dfrac{7}{9} = 12\dfrac{14}{18}$$
$$\overline{\qquad 28\dfrac{41}{18} = 28 + 2\dfrac{5}{18} = 30\dfrac{5}{18}}$$

You Try It 5

Add: $7\dfrac{4}{5} + 6\dfrac{7}{10} + 13\dfrac{11}{15}$

Your solution

Solutions on pp. A6–A7

*O*bjective 2.2D *To subtract proper fractions*

Fractions with the same denominator are subtracted by subtracting the numerators and placing the difference over the common denominator. After subtracting, reduce the fraction to simplest form.

Subtract: $\dfrac{5}{7} - \dfrac{3}{7}$

$$\begin{array}{r} \dfrac{5}{7} \\[2mm] -\dfrac{3}{7} \\[2mm] \hline \dfrac{2}{7} \end{array}$$

Subtract the numerators and place the difference over the common denominator.

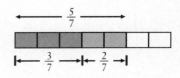

$$\dfrac{5}{7} - \dfrac{3}{7} = \dfrac{5-3}{7} = \dfrac{2}{7}$$

To subtract fractions with different denominators, first rewrite the fractions as equivalent fractions with a common denominator.

Subtract: $\dfrac{5}{6} - \dfrac{1}{4}$

Step 1
Find the LCD of 6 and 4.

LCD = 12

Step 2
Build equivalent fractions using the LCD.

$$\begin{array}{r} \dfrac{5}{6} = \dfrac{10}{12} \\[2mm] -\dfrac{1}{4} = \dfrac{3}{12} \end{array}$$

Step 3
Subtract the fractions.

$$\begin{array}{r} \dfrac{5}{6} = \dfrac{10}{12} \\[2mm] -\dfrac{1}{4} = \dfrac{3}{12} \\[2mm] \hline \dfrac{7}{12} \end{array}$$

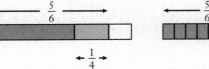

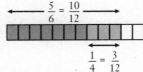

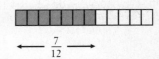

Example 6

Subtract: $\dfrac{11}{16} - \dfrac{5}{12}$

Solution

$$\begin{array}{r} \dfrac{11}{16} = \dfrac{33}{48} \\[2mm] -\dfrac{5}{12} = \dfrac{20}{48} \\[2mm] \hline \dfrac{13}{48} \end{array}$$

The LCD of 16 and 12 is 48.

You Try It 6

Subtract: $\dfrac{13}{18} - \dfrac{7}{24}$

Your solution

Solution on p. A7

*O*bjective 2.2E   *To subtract whole numbers, mixed numbers, and fractions*

To subtract mixed numbers without borrowing, subtract the fractional parts and then subtract the whole numbers.

Subtract: $5\frac{5}{6} - 2\frac{3}{4}$

Step 1
Subtract the fractional parts.

$$5\frac{5}{6} = 5\frac{10}{12}$$
$$-2\frac{3}{4} = 2\frac{9}{12}$$
$$\overline{\phantom{-2\frac{3}{4}=2}\frac{1}{12}}$$

The LCD of 6 and 4 is 12.

Step 2
Subtract the whole numbers.

$$5\frac{5}{6} = 5\frac{10}{12}$$
$$-2\frac{3}{4} = 2\frac{9}{12}$$
$$\overline{\phantom{-2\frac{3}{4}=}3\frac{1}{12}}$$

As with whole numbers, subtraction of mixed numbers sometimes involves borrowing.

Subtract: $5 - 2\frac{5}{8}$

Step 1
Borrow 1 from 5.

$$5 \;= \overset{4}{\cancel{5}}\, 1$$
$$-2\frac{5}{8} = 2\frac{5}{8}$$

Step 2
Write 1 as a fraction so that the fractions have the same denominators.

$$5 \;= 4\frac{8}{8}$$
$$-2\frac{5}{8} = 2\frac{5}{8}$$

Step 3
Subtract the mixed numbers.

$$5 \;= 4\frac{8}{8}$$
$$-2\frac{5}{8} = 2\frac{5}{8}$$
$$\overline{\phantom{-2\frac{5}{8}=}2\frac{3}{8}}$$

Subtract: $7\frac{1}{6} - 2\frac{5}{8}$

Step 1
Build equivalent fractions using the LCD.

$$7\frac{1}{6} = 7\frac{4}{24}$$
$$-2\frac{5}{8} = 2\frac{15}{24}$$

Step 2
Borrow 1 from 7 and add it to $\frac{4}{24}$. Write the result as an improper fraction.

$$7\frac{1}{6} = \overset{6}{\cancel{7}}1\frac{4}{24} = 6\frac{28}{24}$$
$$-2\frac{5}{8} = \;\;2\frac{15}{24} = 2\frac{15}{24}$$

Step 3
Subtract the mixed numbers.

$$7\frac{1}{6} = 6\frac{28}{24}$$
$$-2\frac{5}{8} = 2\frac{15}{24}$$
$$\overline{\phantom{-2\frac{5}{8}=}4\frac{13}{24}}$$

Example 7

Subtract: $15\frac{7}{8} - 12\frac{2}{3}$

Solution

$15\frac{7}{8} = 15\frac{21}{24}$ The LCD of 8 and 3 is 24.

$-12\frac{2}{3} = 12\frac{16}{24}$

$\overline{\phantom{-12\frac{2}{3} = 12}\ 3\frac{5}{24}}$

You Try It 7

Subtract: $17\frac{5}{9} - 11\frac{5}{12}$

Your solution

Example 8

Subtract: $9 - 4\frac{3}{11}$

Solution

$9\phantom{\frac{3}{11}} = 8\frac{11}{11}$

$-4\frac{3}{11} = 4\frac{3}{11}$

$\overline{\phantom{-4\frac{3}{11} = }\ 4\frac{8}{11}}$

You Try It 8

Subtract: $8 - 2\frac{4}{13}$

Your solution

Example 9

Subtract: $11\frac{5}{12} - 2\frac{11}{16}$

Solution

The LCD of 12 and 16 is 48.

$11\frac{5}{12} = 11\frac{20}{48} = 10\frac{68}{48}$

$-\ 2\frac{11}{16} = \ 2\frac{33}{48} = \ 2\frac{33}{48}$

$\overline{\phantom{-\ 2\frac{11}{16} = \ 2\frac{33}{48} = }\ 8\frac{35}{48}}$

You Try It 9

Subtract: $21\frac{7}{9} - 7\frac{11}{12}$

Your solution

Solutions on p. A7

Objective 2.2F   *To solve problems involving shares of stock*

A corporation raises money by selling **shares of stock,** which are certificates that represent ownership in the corporation. Purchasers of shares of stock are called **stockholders.**

The **market price** of a share of stock is determined by how much a buyer is willing to pay for it or for what price a stockholder is willing to sell it. The market prices of stocks are printed daily in major newspapers all over the country. A stock listed at $35\frac{3}{8}$ is selling for $\$35\frac{3}{8}$ per share.

Example 10

Six months ago, a computer company stock was selling at $17\frac{3}{8}$. The market price of the stock has gained $7\frac{3}{4}$ since then. Find the current market price of the stock.

You Try It 10

At the beginning of the week, the Simplex Electric Company stock was selling at $37\frac{1}{4}$. During the week the stock gained $5\frac{3}{8}$. Find the market price of the stock at the end of the week.

Strategy

To find the current market price, add the gain $\left(7\frac{3}{4}\right)$ to the market price six months ago $\left(17\frac{3}{8}\right)$.

Your strategy

Solution

$$17\frac{3}{8} = 17\frac{3}{8}$$
$$+ \ 7\frac{3}{4} = \ 7\frac{6}{8}$$
$$\overline{\rule{2cm}{0.4pt}}$$
$$24\frac{9}{8} = 24 + 1\frac{1}{8} = 25\frac{1}{8}$$

The current market price of the stock is $25\frac{1}{8}$.

Your solution

Solution on p. A7

EXERCISE 2.2A

Find the LCM.

1. 3, 8

2. 5, 7

3. 6, 8

4. 12, 16

5. 8, 14

6. 3, 9

7. 4, 10

8. 14, 42

9. 4, 8, 12

10. 3, 5, 10

11. 5, 12, 18

12. 9, 36, 64

EXERCISE 2.2B

Add.

13. $\dfrac{2}{7} + \dfrac{1}{7}$

14. $\dfrac{3}{8} + \dfrac{7}{8} + \dfrac{1}{8}$

15. $\dfrac{1}{2} + \dfrac{2}{3}$

16. $\dfrac{3}{5} + \dfrac{7}{10}$

17. $\dfrac{1}{6} + \dfrac{7}{9}$

18. $\dfrac{5}{12} + \dfrac{5}{16}$

19. $\dfrac{2}{3} + \dfrac{6}{19}$

20. $\dfrac{1}{3} + \dfrac{5}{6} + \dfrac{7}{9}$

21. $\dfrac{2}{3} + \dfrac{5}{6} + \dfrac{7}{12}$

22. $\dfrac{2}{3} + \dfrac{1}{5} + \dfrac{7}{12}$

23. $\dfrac{2}{3} + \dfrac{3}{5} + \dfrac{7}{8}$

24. $\dfrac{2}{3} + \dfrac{5}{8} + \dfrac{7}{9}$

1. _____
2. _____
3. _____
4. _____
5. _____
6. _____
7. _____
8. _____
9. _____
10. _____
11. _____
12. _____
13. _____
14. _____
15. _____
16. _____
17. _____
18. _____
19. _____
20. _____
21. _____
22. _____
23. _____
24. _____

*E*XERCISE 2.2C

Add.

25. $1\frac{1}{2} + 2\frac{1}{6}$

26. $2\frac{2}{5} + 3\frac{3}{10}$

27. $4\frac{1}{2} + 5\frac{7}{12}$

28. $3\frac{3}{8} + 2\frac{5}{16}$

29. $2\frac{7}{9} + 3\frac{5}{12}$

30. $3\frac{5}{8} + 2\frac{11}{20}$

31. $3\frac{1}{2} + 2\frac{3}{4} + 1\frac{5}{6}$

32. $2\frac{1}{2} + 3\frac{2}{3} + 4\frac{1}{4}$

33. $7\frac{2}{5} + 3\frac{7}{10} + 5\frac{11}{15}$

*E*XERCISE 2.2D

Subtract.

34. $\frac{11}{15} - \frac{3}{15}$

35. $\frac{11}{12} - \frac{7}{12}$

36. $\frac{2}{3} - \frac{1}{6}$

37. $\frac{5}{8} - \frac{2}{7}$

38. $\frac{5}{7} - \frac{3}{14}$

39. $\frac{5}{9} - \frac{7}{15}$

40. $\frac{7}{9} - \frac{1}{6}$

41. $\frac{5}{12} - \frac{5}{16}$

42. $\frac{7}{30} - \frac{3}{20}$

43. $\frac{5}{9} - \frac{1}{12}$

44. $\frac{11}{16} - \frac{5}{12}$

45. $\frac{53}{70} - \frac{13}{35}$

46. $\frac{7}{16} - \frac{5}{24}$

47. $\frac{7}{8} - \frac{5}{9}$

48. $\frac{5}{6} - \frac{3}{5}$

25. _____

26. _____

27. _____

28. _____

29. _____

30. _____

31. _____

32. _____

33. _____

34. _____

35. _____

36. _____

37. _____

38. _____

39. _____

40. _____

41. _____

42. _____

43. _____

44. _____

45. _____

46. _____

47. _____

48. _____

EXERCISE 2.2E

Subtract.

49. $5\dfrac{7}{12} - 2\dfrac{5}{12}$

50. $16\dfrac{11}{15} - 11\dfrac{8}{15}$

51. $6\dfrac{1}{3} - 2$

52. $5\dfrac{7}{8} - 1$

53. $10 - 6\dfrac{1}{3}$

54. $3 - 2\dfrac{5}{9}$

55. $6\dfrac{2}{5} - 4\dfrac{4}{5}$

56. $16\dfrac{3}{8} - 10\dfrac{7}{8}$

57. $23\dfrac{7}{8} - 16\dfrac{2}{3}$

58. $16\dfrac{2}{5} - 8\dfrac{4}{9}$

59. $16\dfrac{3}{10} - 7\dfrac{9}{10}$

60. $14\dfrac{3}{5} - 7\dfrac{8}{9}$

EXERCISE 2.2F

Solve.

61. The market price of a share of stock one year ago was 25. Its market price today is $29\dfrac{3}{8}$. Find the difference between the stock's market price one year ago and its market price today.

62. The market price of a share of stock six months ago was 35. Its market price today is $27\dfrac{3}{8}$. Find the difference between the stock's market price six months ago and its market price today.

63. A stockholder sold 200 shares of stock at $48\dfrac{1}{4}$ per share. The stockholder had earlier purchased the stock at $37\dfrac{3}{8}$ per share. Find the stockholder's gain per share.

64. A stockholder sold 150 shares of stock at $34\dfrac{1}{4}$ per share. The stock had been purchased at $36\dfrac{5}{8}$ per share. Find the stockholder's loss per share.

65. A stock is sold for $2\dfrac{7}{8}$ below its purchase price of 25. Find the stock's market price.

66. Shares of stock are sold for $3\dfrac{1}{4}$ above their purchase price of $50\dfrac{1}{2}$. Find the stock's market price.

49. _____

50. _____

51. _____

52. _____

53. _____

54. _____

55. _____

56. _____

57. _____

58. _____

59. _____

60. _____

61. _____

62. _____

63. _____

64. _____

65. _____

66. _____

67. One day on the stock exchange, the highest price paid for one share of a company's stock was $29\frac{3}{4}$, and the lowest price paid was $28\frac{7}{8}$. Find the difference between the high and low market prices per share that day.

68. A stockholder sold 100 shares of stock that had been purchased at $83\frac{5}{8}$ per share. The stock had gained $5\frac{1}{2}$ per share. Find the market price per share.

69. At the beginning of the year, the stock in a cellular phone company was selling for $26\frac{5}{8}$ per share. The price of the stock gained $13\frac{3}{4}$ per share during a six-month period. Find the price of the stock at the end of the six months.

70. At the beginning of the week, an airline's stock was selling at $158\frac{3}{8}$ per share. During the week, the stock gained $28\frac{3}{4}$ per share. Find the price of the stock at the end of the week.

71. A gas-and-electric stock was purchased at $26\frac{3}{4}$ per share. Monthly gains for the first three months of ownership were $1\frac{1}{2}$, $2\frac{5}{8}$, and $\frac{1}{4}$. Find the market price of the stock at the end of the three months.

72. A telephone company stock was bought at $55\frac{3}{8}$ per share. Monthly gains for the first three months of ownership were $\frac{3}{8}$, $1\frac{1}{2}$, and $\frac{3}{4}$. Find the market price of the stock at the end of the three months.

73. An investor bought 100 shares of a utility stock at $24\frac{1}{2}$. The stock gained $5\frac{5}{8}$ during the first month of ownership and lost $2\frac{1}{4}$ during the second month. Find the value of 1 share of the utility stock at the end of the second month.

74. An automobile company stock was purchased at $37\frac{1}{2}$ per share. During the first month of ownership, the stock gained $2\frac{3}{8}$ per share. During the second month of ownership, the stock lost $1\frac{1}{4}$ per share. Find the market price of the stock at the end of the two months.

75. An aircraft company stock was bought at $45\frac{5}{8}$ per share. During the first month of ownership, the stock dropped $2\frac{3}{4}$ per share. During the second month of ownership, the stock gained $1\frac{1}{2}$ per share. Find the market price of the stock at the end of the two months.

67. _____

68. _____

69. _____

70. _____

71. _____

72. _____

73. _____

74. _____

75. _____

 **S**ECTION 2.3 | ## Multiplication and Division of Fractions and Mixed Numbers

Objective 2.3A   *To multiply proper fractions*

The product of two fractions is the product of the numerators over the product of the denominators.

Multiply: $\dfrac{2}{3} \times \dfrac{4}{5}$

Multiply the numerators. $\dfrac{2}{3} \times \dfrac{4}{5} = \dfrac{2 \cdot 4}{3 \cdot 5} = \dfrac{8}{15}$
Multiply the denominators.

The product $\dfrac{2}{3} \times \dfrac{4}{5}$ can be read "$\dfrac{2}{3}$ times $\dfrac{4}{5}$" or "$\dfrac{2}{3}$ of $\dfrac{4}{5}$."

Reading the times sign as "of" is useful in application problems involving fractions and in diagramming the product of fractions.

Here $\dfrac{4}{5}$ of the bar is shaded.

Shade $\dfrac{2}{3}$ of the $\dfrac{4}{5}$ already shaded.

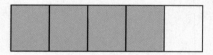

$\dfrac{8}{15}$ of the bar is then shaded.

$\dfrac{2}{3}$ of $\dfrac{4}{5} = \dfrac{8}{15}$

After multiplying two fractions, reduce the product to simplest form.

Multiply: $\dfrac{3}{4} \times \dfrac{14}{15}$

$$\dfrac{3}{4} \times \dfrac{14}{15} = \dfrac{3 \cdot 14}{4 \cdot 15} = \dfrac{42}{60} = \dfrac{42 \div 6}{60 \div 6} = \dfrac{7}{10}$$

Example 1

Multiply: $\dfrac{4}{5} \times \dfrac{3}{8}$

Solution

$\dfrac{4}{5} \times \dfrac{3}{8} = \dfrac{4 \cdot 3}{5 \cdot 8} = \dfrac{12}{40} = \dfrac{12 \div 4}{40 \div 4} = \dfrac{3}{10}$

You Try It 1

Multiply: $\dfrac{4}{9} \times \dfrac{3}{10}$

Your solution

Solution on p. A8

Objective 2.3B *To multiply whole numbers, mixed numbers, and fractions*

To multiply a whole number by a fraction or a mixed number, first write the whole number as a fraction with a denominator of 1.

Multiply: $4 \times \dfrac{3}{7}$

$$4 \times \frac{3}{7} = \frac{4}{1} \times \frac{3}{7} = \frac{4 \cdot 3}{1 \cdot 7} = \frac{12}{7} = 1\frac{5}{7}$$

When one of the factors in a product is a mixed number, write the mixed number as an improper fraction before multiplying.

Multiply: $2\dfrac{1}{3} \times \dfrac{3}{14}$

$$2\frac{1}{3} \times \frac{3}{14} = \frac{7}{3} \times \frac{3}{14} = \frac{7 \cdot 3}{3 \cdot 14} = \frac{21}{42} = \frac{21 \div 21}{42 \div 21} = \frac{1}{2}$$

Multiply: $5\dfrac{3}{4} \times 2\dfrac{1}{2}$

$$5\frac{3}{4} \times 2\frac{1}{2} = \frac{23}{4} \times \frac{5}{2} = \frac{23 \cdot 5}{4 \cdot 2} = \frac{115}{8} = 14\frac{3}{8}$$

Example 2

Multiply: $4\dfrac{2}{5} \times 7$

Solution

$$4\frac{2}{5} \times 7 = \frac{22}{5} \times \frac{7}{1} = \frac{22 \cdot 7}{5 \cdot 1} = \frac{154}{5} = 30\frac{4}{5}$$

You Try It 2

Multiply: $3\dfrac{2}{7} \times 6$

Your solution

Solution on p. A8

Objective 2.3C *To divide proper fractions*

The **reciprocal** of a fraction is the fraction with the numerator and denominator interchanged. The process of interchanging the numerator and denominator of a fraction is called **inverting.**

The reciprocal of $\dfrac{2}{3}$ is $\dfrac{3}{2}$.

To find the reciprocal of a whole number, first write the whole number as a fraction with a denominator of 1; then find the reciprocal of that fraction.

The reciprocal of 5 is $\dfrac{1}{5}$. $\left(\text{Think } 5 = \dfrac{5}{1}\right)$

Reciprocals are used to rewrite division problems as related multiplication problems. Look at the following two problems.

$$8 \div 2 = 4 \qquad\qquad 8 \times \frac{1}{2} = 4$$

8 divided by 2 is 4.   8 times the reciprocal of 2 is 4.

"Divided by" means the same thing as "times the reciprocal of." So "÷ 2" can be replaced with "× $\frac{1}{2}$" and the answer will be the same. Fractions are divided by making this replacement.

Divide: $\frac{2}{3} \div \frac{3}{4}$

Rewrite division by $\frac{3}{4}$ as multiplication by the reciprocal of $\frac{3}{4}$, which is $\frac{4}{3}$. Then multiply the fractions.

$$\frac{2}{3} \div \frac{3}{4} = \frac{2}{3} \times \frac{4}{3} = \frac{2 \cdot 4}{3 \cdot 3} = \frac{8}{9}$$

Example 3

Divide: $\frac{3}{5} \div \frac{12}{25}$

Solution

$$\frac{3}{5} \div \frac{12}{25} = \frac{3}{5} \times \frac{25}{12} = \frac{3 \cdot 25}{5 \cdot 12} = \frac{75}{60}$$

$$= \frac{75 \div 15}{60 \div 15} = \frac{5}{4} = 1\frac{1}{4}$$

You Try It 3

Divide: $\frac{3}{4} \div \frac{9}{10}$

Your solution

Solution on p. A8

Objective 2.3D   *To divide whole numbers, mixed numbers, and fractions*

To divide a fraction and a whole number, first write the whole number as a fraction with a denominator of 1.

Divide: $\frac{3}{7} \div 5$

$5 = \frac{5}{1}$. Rewrite division by $\frac{5}{1}$ as multiplication by the reciprocal of $\frac{5}{1}$, which is $\frac{1}{5}$. Then multiply the fractions.

$$\frac{3}{7} \div 5 = \frac{3}{7} \div \frac{5}{1} = \frac{3}{7} \times \frac{1}{5} = \frac{3 \cdot 1}{7 \cdot 5} = \frac{3}{35}$$

When one of the numbers in a quotient is a mixed number, write the mixed number as an improper fraction before dividing.

Divide: $2\frac{2}{3} \div \frac{3}{4}$

$$2\frac{2}{3} \div \frac{3}{4} = \frac{8}{3} \div \frac{3}{4} = \frac{8}{3} \times \frac{4}{3} = \frac{8 \cdot 4}{3 \cdot 3} = \frac{32}{9} = 3\frac{5}{9}$$

Example 4

Divide: $2\dfrac{3}{4} \div 1\dfrac{5}{7}$

Solution

$$2\dfrac{3}{4} \div 1\dfrac{5}{7} = \dfrac{11}{4} \div \dfrac{12}{7} = \dfrac{11}{4} \times \dfrac{7}{12}$$

$$= \dfrac{11 \cdot 7}{4 \cdot 12} = \dfrac{77}{48} = 1\dfrac{29}{48}$$

You Try It 4

Divide: $3\dfrac{2}{3} \div 2\dfrac{2}{5}$

Your solution

Solution on p. A8

Objective 2.3E *To solve problems involving hourly wage*

Many workers are paid on the basis of the number of hours they work. These employees receive an **hourly wage** for each hour worked. Usually employees who earn an hourly wage are paid a regular hourly rate for working 40 hours a week and receive a higher rate for **overtime,** the time worked in excess of 40 hours. For each hour of overtime, an employee commonly receives $1\dfrac{1}{2}$ times the regular hourly rate. This is referred to as "time and a half."

An employee who earns time and a half for overtime and whose regular hourly rate is $8 per hour will earn $12 for each hour of overtime worked.

$$8 \times 1\dfrac{1}{2} = \dfrac{8}{1} \times \dfrac{3}{2} = 12$$

Example 5

This week a nurse worked 6 hours of overtime at time and a half. If the nurse's regular hourly wage is $14, find the nurse's overtime pay for this week.

Strategy

To find the nurse's overtime pay:

* Find the overtime rate by multiplying the hourly wage ($14) by $1\dfrac{1}{2}$.

* Multiply the overtime rate by the number of hours of overtime worked (6).

Solution

$$14 \times 1\dfrac{1}{2} = \dfrac{14}{1} \times \dfrac{3}{2} = 21$$

$$21 \times 6 = 126$$

The nurse's overtime pay for this week is $126.

You Try It 5

Shannon O'Hara is paid time and a half for working overtime. Shannon earned $162 for working 9 hours of overtime this week. Find Shannon's regular hourly wage.

Your strategy

Your solution

Solution on p. A8

*E*XERCISE 2.3A

Multiply.

1. $\frac{2}{3} \times \frac{7}{8}$

2. $\frac{1}{2} \times \frac{2}{3}$

3. $\frac{1}{2} \times \frac{5}{6}$

4. $\frac{1}{6} \times \frac{1}{8}$

5. $\frac{11}{12} \times \frac{3}{5}$

6. $\frac{2}{5} \times \frac{4}{9}$

7. $\frac{1}{5} \times \frac{5}{8}$

8. $\frac{3}{5} \times \frac{3}{10}$

9. $\frac{5}{6} \times \frac{1}{2}$

10. $\frac{7}{8} \times \frac{3}{14}$

11. $\frac{5}{12} \times \frac{6}{7}$

12. $\frac{12}{5} \times \frac{5}{8}$

*E*XERCISE 2.3B

Multiply.

13. $14 \times \frac{5}{7}$

14. $\frac{2}{3} \times 6$

15. $\frac{1}{3} \times 1\frac{1}{3}$

16. $\frac{2}{5} \times 2\frac{1}{2}$

17. $4 \times 2\frac{1}{2}$

18. $9 \times 3\frac{1}{3}$

19. $\frac{3}{8} \times 4\frac{4}{5}$

20. $\frac{3}{8} \times 4\frac{1}{2}$

21. $1\frac{1}{3} \times 2\frac{1}{4}$

22. $1\frac{1}{2} \times 5\frac{1}{2}$

23. $2\frac{2}{5} \times 1\frac{7}{12}$

24. $3\frac{1}{3} \times 6\frac{3}{5}$

1. _____

2. _____

3. _____

4. _____

5. _____

6. _____

7. _____

8. _____

9. _____

10. _____

11. _____

12. _____

13. _____

14. _____

15. _____

16. _____

17. _____

18. _____

19. _____

20. _____

21. _____

22. _____

23. _____

24. _____

25. $5\frac{3}{7} \times 3\frac{1}{2}$ **26.** $2\frac{2}{7} \times 3\frac{1}{2}$ **27.** $2\frac{2}{3} \times 3\frac{1}{4}$

28. $5\frac{1}{5} \times 3\frac{3}{4}$ **29.** $\frac{5}{8} \times 5\frac{3}{5}$ **30.** $3\frac{1}{5} \times 2\frac{2}{3}$

EXERCISE 2.3C

Divide.

31. $\frac{1}{3} \div \frac{2}{5}$ **32.** $\frac{3}{7} \div \frac{3}{2}$ **33.** $\frac{3}{7} \div \frac{3}{7}$

34. $\frac{8}{9} \div \frac{4}{5}$ **35.** $\frac{2}{5} \div \frac{4}{7}$ **36.** $\frac{1}{2} \div \frac{1}{4}$

37. $\frac{1}{3} \div \frac{1}{9}$ **38.** $\frac{5}{8} \div \frac{15}{2}$ **39.** $\frac{5}{16} \div \frac{3}{8}$

40. $\frac{5}{7} \div \frac{2}{7}$ **41.** $\frac{5}{6} \div \frac{1}{9}$ **42.** $\frac{15}{8} \div \frac{5}{32}$

EXERCISE 2.3D

Divide.

43. $\frac{5}{6} \div 25$ **44.** $22 \div \frac{3}{11}$ **45.** $6 \div 3\frac{1}{3}$

46. $5\frac{1}{2} \div 11$ **47.** $3\frac{1}{3} \div \frac{3}{8}$ **48.** $6\frac{1}{2} \div \frac{1}{2}$

25. _____
26. _____
27. _____
28. _____
29. _____
30. _____
31. _____
32. _____
33. _____
34. _____
35. _____
36. _____
37. _____
38. _____
39. _____
40. _____
41. _____
42. _____
43. _____
44. _____
45. _____
46. _____
47. _____
48. _____

49. $\dfrac{3}{8} \div 2\dfrac{1}{4}$ **50.** $\dfrac{5}{12} \div 4\dfrac{4}{5}$ **51.** $1\dfrac{1}{2} \div 1\dfrac{3}{8}$

52. $2\dfrac{1}{4} \div 1\dfrac{3}{8}$ **53.** $1\dfrac{3}{5} \div 2\dfrac{1}{10}$ **54.** $2\dfrac{5}{6} \div 1\dfrac{1}{9}$

55. $2\dfrac{1}{3} \div 3\dfrac{2}{3}$ **56.** $4\dfrac{1}{2} \div 2\dfrac{1}{6}$ **57.** $7\dfrac{1}{2} \div 2\dfrac{2}{3}$

58. $8\dfrac{1}{4} \div 2\dfrac{3}{4}$ **59.** $6\dfrac{1}{3} \div 5$ **60.** $8\dfrac{2}{7} \div 1$

49. _____

50. _____

51. _____

52. _____

53. _____

54. _____

55. _____

56. _____

57. _____

58. _____

59. _____

60. _____

61. _____

62. _____

63. _____

64. _____

65. _____

66. _____

*E*XERCISE 2.3E

Solve.

61. A dental hygienist receives time and a half for working overtime. The hygienist's regular hourly rate is $16. Find the hygienist's overtime rate.

62. An electrician receives time and a half for working over 40 hours per week. The electrician's regular hourly rate is $22. Find the electrician's overtime rate.

63. This week a carpenter worked 5 hours of overtime at time and a half. The carpenter's regular hourly wage is $18. Find the carpenter's overtime pay for this week.

64. On Saturday a plumber worked 4 hours at time and a half. The plumber's regular hourly wage is $20. Find the plumber's overtime pay for Saturday.

65. Jon McCloud receives time and a half for working overtime and is paid $15 for each hour of overtime worked. Find Jon's regular hourly wage.

66. William Carey receives time and a half for working overtime. If William is paid $42 for each hour of overtime worked, what is William's regular hourly wage?

67. Jeanna Carrera is paid time and a half for working overtime. Jeanna earned $96 for working 8 hours of overtime this week. Find Jeanna's regular hourly wage.

68. Rosalinda Johnson, who is paid time and a half for working overtime, earned $126 for working 7 hours of overtime this week. Find Rosalinda's regular hourly rate.

69. An employee whose regular hourly wage is $8 per hour worked 4 hours of overtime on Friday and $5\frac{1}{2}$ hours of overtime on Saturday. The employee is paid time and a half for overtime. Find the employee's overtime pay.

70. A typist earns $6 per hour and time and a half for working overtime. Find the typist's earnings for working overtime from 8 A.M. until noon on Saturday and from 2 P.M. until 4 P.M. on Sunday.

71. An employee who is paid time and a half for time over 40 hours worked $42\frac{1}{2}$ hours this week. The employee earns $12 per hour. Find the employee's earnings for the week.

72. Kristi Yang is paid an hourly wage of $8 per hour and time and a half for time over 40 hours. Find Kristi's earnings for working $48\frac{1}{4}$ hours this week.

73. Hector Elizondo receives a regular hourly wage of $16 per hour and time and a half for time over 40 hours. Find Hector's earnings for working $46\frac{1}{2}$ hours this week.

74. This week a computer operator worked $8\frac{1}{4}$ hours on Monday, $9\frac{1}{2}$ hours on Tuesday, $7\frac{3}{4}$ hours on Wednesday, $8\frac{1}{2}$ hours on Thursday, and $8\frac{1}{2}$ hours on Friday. The computer operator is paid $20 per hour and time and a half for time over 40 hours worked in the week. Find the computer operator's earnings for this week's work.

75. During the past week, a machinist worked $9\frac{1}{4}$ hours on Monday, $8\frac{3}{4}$ hours on Tuesday, 7 hours on Wednesday, $9\frac{1}{4}$ hours on Thursday, and $8\frac{3}{4}$ hours on Friday. The machinist is paid $14 per hour and time and a half for time over 40 hours. Find the machinist's earnings for the past week.

67. _____

68. _____

69. _____

70. _____

71. _____

72. _____

73. _____

74. _____

75. _____

TIME CARDS

An employee who is paid an hourly wage is usually required to keep a timecard. A weekly timecard shows the times worked each day of the week. The hours worked each day are computed, and these hours are added to find the total number of hours worked during the week.

Many of the time clocks that record the time an employee arrives at work and leaves the office use the 24-hour clock. On a 24-hour clock, the hours from 1 A.M. until noon are the same as on the 12-hour clock. After the hour of noon, the hours are numbered sequentially from 12 o'clock.

12-Hour Clock	24-Hour Clock
11:00 A.M.	11:00
12:00 noon	12:00
1:00 P.M.	13:00
2:00 P.M.	14:00
3:00 P.M.	15:00
4:00 P.M.	16:00
5:00 P.M.	17:00
6:00 P.M.	18:00
.	.
.	.
.	.
12:00 midnight	24:00

If the time on a 24-hour clock is 12:59 or earlier, the hour is the same as the A.M. time on a 12-hour clock. If the time on a 24-hour clock is 13:00 or later, subtract 12 from the hour to find the P.M. time on a 12-hour clock.

A 24-hour clock reads 19:32. What is the time on a 12-hour clock?

> The time on the 24-hour clock is after 12:59.
> Subtract 12 from the hours (19). $19 - 12 = 7$
> The number of minutes (32) remains unchanged.
> The time is 7:32 P.M.

The hours worked each day are usually rounded to the nearest quarter-hour.

What fractional part of an hour is 45 minutes?

> Write a fraction with 45 in the numerator and the
> number of minutes in one hour (60) in the denominator. $\dfrac{45}{60} = \dfrac{3}{4}$
> Reduce the fraction to simplest form.
> 45 minutes is $\frac{3}{4}$ of an hour.

The method used in the example above can be used to show that 15 minutes is $\frac{1}{4}$ of an hour, and that 30 minutes is $\frac{1}{2}$ of an hour. Hours worked each day are rounded to the nearest $\frac{1}{4}$, $\frac{1}{2}$, or $\frac{3}{4}$ of an hour.

Round 3 hours 22 minutes to the nearest quarter-hour.

15 minutes is $\frac{1}{4}$ hour.
Find the difference between 22 minutes and 15 minutes. $22 - 15 = 7$

30 minutes is $\frac{1}{2}$ hour.
Find the difference between 30 minutes and 22 minutes. $30 - 22 = 8$

Because 7 is less than 8, 22 minutes is closer to $\frac{1}{4}$ hour than to $\frac{1}{2}$ hour.

3 hours 22 minutes rounded to the nearest quarter-hour is $3\frac{1}{4}$ hour.

Find the number of hours between 12:57 and 16:43. Round to the nearest quarter-hour.

43 is less than 57.
Borrowing is necessary.

$16:43 = 16$ hours 43 minutes
$-12:57 = 12$ hours 57 minutes

1 hour = 60 minutes
Borrow 1 hour from 16 hours and
add 60 minutes to 43 minutes.

15 hours 103 minutes
-12 hours   57 minutes

Subtract the hours. Subtract the minutes.   3 hours 46 minutes

3 hours 46 minutes is approximately $3\frac{3}{4}$ hours.

Complete the weekly timecards. Round the hours worked each day to the nearest quarter-hour.

TEMPORARY EMPLOYEE TIMECARD			NAME: ▁▁▁▁▁▁▁▁▁▁▁ DEPT: Data Processing		
Date	In	Out	In	Out	Hours
7/16	8:29	12:01	12:35	16:41	1.
7/17	8:26	12:05	12:31	16:38	2.
7/18	8:31	12:16	12:52	16:28	3.
7/19	8:27	12:30	13:04	16:52	4.
7/20	8:32	12:15	12:47	16:36	5.
				Total Hours	6.
Employee Signature ▁▁▁▁▁▁▁▁			Supervisor Signature ▁▁▁▁▁▁▁▁		

TEMPORARY EMPLOYEE TIMECARD			NAME: ▁▁▁▁▁▁▁▁▁▁▁ DEPT: Maintenance		
Date	In	Out	In	Out	Hours
9/22	9:00	12:19	12:43	16:47	7.
9/23	8:46	12:00	12:34	16:49	8.
9/24	9:17	11:45	12:17	16:10	9.
9/25	9:10	12:06	12:35	16:28	10.
9/26	8:53	13:09	13:36	15:42	11.
				Total Hours	12.
Employee Signature ▁▁▁▁▁▁▁▁			Supervisor Signature ▁▁▁▁▁▁▁▁		

1. ▁▁▁▁▁▁▁▁▁▁▁▁

2. ▁▁▁▁▁▁▁▁▁▁▁▁

3. ▁▁▁▁▁▁▁▁▁▁▁▁

4. ▁▁▁▁▁▁▁▁▁▁▁▁

5. ▁▁▁▁▁▁▁▁▁▁▁▁

6. ▁▁▁▁▁▁▁▁▁▁▁▁

7. ▁▁▁▁▁▁▁▁▁▁▁▁

8. ▁▁▁▁▁▁▁▁▁▁▁▁

9. ▁▁▁▁▁▁▁▁▁▁▁▁

10. ▁▁▁▁▁▁▁▁▁▁▁▁

11. ▁▁▁▁▁▁▁▁▁▁▁▁

12. ▁▁▁▁▁▁▁▁▁▁▁▁

BUSINESS CASE STUDY

Real Estate   You are interested in purchasing land on which to build storage facilities. You have researched the need for rental facilities in your area and have determined that you should build a one-story building with 40 storage units. The floor space of each unit will measure 30 feet by 50 feet. In addition to this building, you plan on using a total of $\frac{3}{4}$ acre of the property you purchase for driveways, for a small office building, and for distance from your buildings to the abutting properties.

You have looked at two different pieces of property. One is listed at $31,800 and is $2\frac{1}{2}$ acres. The second lot, listed at $39,975, is $3\frac{1}{4}$ acres. Both lots are commercially zoned. (A **commercial zone** is an area that is designated for businesses.)

1. Is either lot adequate for your building needs? (One acre is equal to 43,560 square feet).
2. **a.** What is the asking price per acre of the $2\frac{1}{2}$-acre lot?
   b. What is the asking price per acre of the $3\frac{1}{4}$-acre lot?
   c. What is the difference between the asking prices per acre?
3. Is price per acre a consideration in deciding which property to purchase? What other considerations should be taken into account?

CHAPTER SUMMARY

Key Words   A **fraction** can represent the number of equal parts of a whole. In a fraction, the **fraction bar** separates the **numerator** above from the **denominator** below. (Objective 2.1A)

In a **proper fraction,** the numerator is smaller than the denominator; a proper fraction is a number less than 1. In an **improper fraction,** the numerator is greater than or equal to the denominator; an improper fraction is a number greater than or equal to 1. A **mixed number** is a number greater than 1 with a whole-number part and a fractional part. (Objective 2.1A)

Equal fractions with different denominators are called **equivalent fractions.** (Objective 2.1B)

A whole-number **factor** of a number divides that number evenly (there is no remainder). A number that is a factor of two or more numbers is a **common factor** of those numbers. (Objective 2.1C)

A fraction is in **simplest form** when the numerator and denominator have no common factors other than 1. (Objective 2.1C)

The **multiples** of a number are the products of that number and the numbers 1, 2, 3, 4, 5, and so on. A number that is a multiple of two or more numbers is a **common multiple** of those numbers. The **least common multiple** (LCM) is the smallest common multiple of two or more numbers. (Objective 2.2A)

The least common multiple of the numbers in the denominators of two or more fractions is called the **lowest common denominator** (LCD). (Objective 2.2B)

A number is a **prime number** if its only whole-number factors are 1 and itself. (Objective 2.2A)

The **reciprocal** of a fraction is the fraction with the numerator and denominator interchanged. The process of interchanging the numerator and denominator of a fraction is called **inverting.** (Objective 2.3C)

Shares of stock are certificates that represent ownership in a corporation. **Stockholders** own shares of stock in a corporation. The **market price** of a share of stock is determined by how much a buyer is willing to pay for it or for what price a stockholder is willing to sell it. (Objective 2.2F)

Employees paid on the basis of the number of hours they work receive an **hourly wage. Overtime** is the time worked in excess of a standard day (usually 8 hours) or a standard week (usually 40 hours). (Objective 2.3E)

Essential Rules

To write an improper fraction as a mixed number, divide the numerator by the denominator; this is the whole-number part of the mixed number. To write the fractional part of the mixed number, write the remainder over the divisor. (Objective 2.1A)

To write a mixed number as an improper fraction, multiply the denominator of the fractional part by the whole-number part. The sum of this product and the numerator of the fractional part is the numerator of the improper fraction. The denominator remains the same. (Objective 2.1A)

To build an equivalent fraction, multiply the numerator and denominator by the same number. (Objective 2.1B)

To reduce a fraction to simplest form, divide the numerator and denominator by a common factor so their only common factor is 1. (Objective 2.1C)

To add fractions with the same denominators, add the numerators and place the sum over the common denominator. (Objective 2.2B)

To add fractions with different denominators, first rewrite the fractions as equivalent fractions with a common denominator. Then add the fractions. (Objective 2.2B)

To subtract fractions with the same denominators, subtract the numerators and place the difference over the common denominator. (Objective 2.2D)

To subtract fractions with different denominators, first rewrite the fractions as equivalent fractions with a common denominator. Then subtract the fractions. (Objective 2.2D)

To multiply two fractions, multiply the numerators; this is the numerator of the product. Then multiply the denominators; this is the denominator of the product. (Objective 2.3A)

To divide two fractions, multiply the first fraction by the reciprocal of the second fraction. (Objective 2.3C)

Divisibility Rules

A number can be divided by 2 if the ones' digit is 0, 2, 4, 6, or 8.
A number can be divided by 3 if the sum of the digits is divisible by 3.
A number can be divided by 4 if the last two digits are divisible by 4.
A number can be divided by 5 if the ones' digit is 0 or 5.
A number can be divided by 9 if the sum of the digits is divisible by 9.
A number can be divided by 10 if the ones' digit is 0.

REVIEW / TEST

1. Write $\frac{18}{5}$ as a mixed number.

2. Write $9\frac{4}{5}$ as an improper fraction.

3. Build an equivalent fraction with the given denominator.

   $\frac{5}{8} = \frac{}{32}$

4. Reduce $\frac{20}{32}$ to simplest form.

5. Find the LCM of 10 and 12.

6. Add: $\frac{7}{12} + \frac{11}{12} + \frac{1}{12}$

7. Add: $\frac{5}{6} + \frac{7}{9}$

8. Add: $2\frac{3}{4} + 5\frac{1}{6}$

9. Subtract: $\frac{17}{24} - \frac{11}{24}$

10. Subtract: $\frac{7}{8} - \frac{3}{4}$

1. _____
2. _____
3. _____
4. _____
5. _____
6. _____
7. _____
8. _____
9. _____
10. _____

11. Subtract: $23 - 9\frac{1}{8}$

12. Multiply: $\frac{9}{10} \times \frac{4}{7}$

13. Multiply: $2\frac{2}{3} \times 1\frac{3}{4}$

14. Divide: $\frac{5}{9} \div \frac{3}{8}$

15. Divide: $3 \div 4\frac{1}{5}$

16. Divide: $3\frac{2}{3} \div 3\frac{1}{6}$

11. _____

12. _____

13. _____

14. _____

15. _____

16. _____

17. _____

18. _____

19. _____

20. _____

17. The market price of a share of stock one year ago was 50. Its market price today is $37\frac{3}{8}$. Find the difference between the stock's market price one year ago and its market price today.

18. Keystone Oil Company stock was purchased at $42\frac{3}{8}$ per share. Monthly gains for the first two months of ownership were $\frac{5}{8}$ and $1\frac{1}{2}$. Find the value of the stock at the end of the two months.

19. An employee who is paid time and a half for working overtime earned $60 for working 5 hours of overtime this week. Find the employee's regular hourly rate.

20. Ed Pabas is paid $12 per hour and time and a half for time over 40 hours. Find Ed's earnings for working $44\frac{1}{2}$ hours this week.

3

Decimals and Business Applications

OBJECTIVES

3.1A To write decimals in standard form and in words

3.1B To add and subtract decimals

3.1C To complete an expense form

3.2A To round a decimal to a given place value

3.2B To multiply decimals

3.2C To divide decimals

3.2D To convert fractions to decimals and decimals to fractions

3.2E To complete a purchase order

3.3A To convert units of length, mass, and capacity in the metric system of measurement

3.3B To calculate freight charges

A S A CONSUMER, you are already familiar with decimals. For example, $5.98 is in decimal notation, and you know that this means five dollars and ninety-eight cents. The decimal part of the number represents a number less than one; $ .98 is less than one dollar. The decimal point (.) separates the whole-number part from the decimal part.

One dollar is equal to 100 cents. One cent, or $ .01, is 1/100 of a dollar. The decimal 0.01 is equal to the fraction 1/100. One of the lessons in this chapter describes how to write a fraction as a decimal and how to write a decimal as a fraction.

You know that $62 and $62.00 both represent sixty-two dollars. A whole number can be written as a decimal by writing a decimal point to the right of the last digit. For example,

$$62 = 62. \qquad 497 = 497.$$

Any number of zeros may be written to the right of the decimal point in a whole number without changing the value of the number.

$$62 = 62.00 = 62.0000 \qquad 497 = 497.0 = 497.000$$

Also, any number of zeros may be written to the right of the last digit in a decimal without changing the value of the number.

$$0.8 = 0.80 = 0.800$$

$$1.35 = 1.350 = 1.3500 = 1.35000 = 1.350000$$

You also know that $ .24 is less than $ .42 because 24¢ is less than 42¢. So you already have a concept of the relative order of two decimals.

In this chapter, addition, subtraction, multiplication, and division with decimals is presented. The metric system of measurement is introduced in this chapter because it is a decimal system. Among the business applications of these concepts are completing expense forms, completing purchase orders, and calculating freight charges. Each of these applications is discussed in this chapter.

Another application of operations with decimals is in the area of banking, which is the topic of Chapter 4. In that chapter, you will use the operations of addition and subtraction of decimals to calculate checkbook balances and to reconcile bank statements.

70

 **S**ECTION 3.1   ## Addition and Subtraction of Decimals

*O*bjective 3.1A   *To write decimals in standard form and in words*

The United States Postal Service requires that a postal card measure at least 0.007 inch thick. The number 0.007 is in **decimal notation.**

Note the relationship between fractions and numbers written in decimal notation.

Three tenths   Three hundredths   Three thousandths

$\dfrac{3}{10} = 0.\underline{3}$ $\dfrac{3}{100} = 0.\underline{03}$ $\dfrac{3}{1000} = 0.\underline{003}$

1 zero   1 decimal place   2 zeros   2 decimal places   3 zeros   3 decimal places

A number written in decimal notation has three parts.	351	.	7089

A number written in decimal notation has three parts.   351   .   7089
Whole-number part   **Decimal point**   **Decimal part**

A number written in decimal notation is often called simply a **decimal.** The position of a digit in a decimal determines the digit's place value.

In the decimal 351.7089, the position of the digit 9 determines that its place value is ten-thousandths.

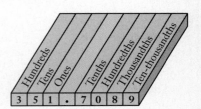

When writing a decimal in words, write the decimal part as though it were a whole number. Then name the place value of the last digit.

0.6481   Six thousand four hundred eighty-one ten-thousandths

549.238   Five hundred forty-nine and two hundred thirty-eight thousandths
(The decimal point is read as "and.")

In business, decimals are often used to represent large whole numbers. For example, a business will report a profit of $1.3 million.

To write the number 1.3 million in standard form, move the decimal point to the right so that the number to the left of the decimal point is in the given place-value position.

The 1 must be in the millions' place.
Move the decimal point 6 places to the right.   1.3 million = 1,300,000

To write a decimal in standard form, you may have to insert zeros after the decimal point so that the last digit is in the given place-value position.

Five and thirty-eight <u>hundredths</u>

    The 8 is in the hundredths' place.   5.3<u>8</u>

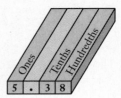

Nineteen and four <u>thousandths</u>

    The 4 is in the thousandths' place.   19.00<u>4</u>
    Insert zeros.

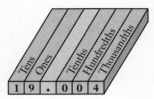

Seventy-one <u>ten-thousandths</u>

    The 1 is in the ten-thousandths'   0.007<u>1</u>
    place. Insert zeros.

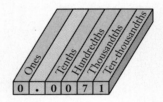

Example 1

Write 307.4027 in words.

Solution

three hundred seven and four thousand twenty-seven ten-thousandths

Example 2

Write 4.58 trillion in standard form.

Solution

4,580,000,000,000

Example 3

Write six hundred seven and seven hundred eight hundred-thousandths in standard form.

Solution

607.00708

You Try It 1

Write 209.05838 in words.

Your solution

You Try It 2

Write 27.6 billion in standard form.

Your solution

You Try It 3

Write forty-two thousand and two hundred seven thousandths in standard form.

Your solution

Solutions on p. A9

***O*bjective 3.1B**　　*To add and subtract decimals*

To add decimals, write the numbers so that the decimal points are lined up vertically. Add as for whole numbers, and write the decimal point in the sum directly below the decimal points in the addends.

Add:　0.237 + 4.9 + 27.32

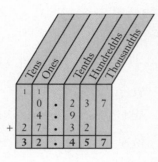

Aligning the decimal points vertically ensures that digits of the same place value are added.

Add:　6.5 + 1.94 + 70.38

$$
\begin{array}{r}
\scriptstyle 1\ 1 \\
6.5 \\
1.94 \\
+\ 70.38 \\
\hline
78.82
\end{array}
$$

To subtract decimals, write the numbers so that the decimal points are lined up vertically. Subtract as for whole numbers, and write the decimal point in the difference directly below the decimal point in the subtrahend.

Subtract 21.532 − 9.875 and check.

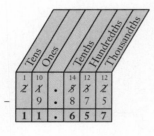

Aligning the decimal points vertically ensures that digits of the same place value are subtracted.

$$
\begin{array}{llr}
 & & \scriptstyle 1\ 1\ 1\ 1 \\
\text{Check:} & \text{Subtrahend} & 9.875 \\
 & +\ \text{Difference} & +\ 11.657 \\
 & =\ \text{Minuend} & 21.532
\end{array}
$$

Subtract 4.3 − 1.7942 and check.

Insert zeros in the minuend before subtracting.
4.3 = 4.3000

$$
\begin{array}{r}
\scriptstyle 3\ \ 12\ 9\ 9\ 10 \\
\cancel{4}.\ \cancel{3}\ \cancel{0}\ \cancel{0}\ \cancel{0} \\
-\ 1.\ 7\ 9\ 4\ 2 \\
\hline
2.\ 5\ 0\ 5\ 8
\end{array}
$$

$$
\begin{array}{lr}
 & \scriptstyle 1\ \ 1\ \ 1\ \ 1 \\
\text{Check:} & 1.\ 7\ 9\ 4\ 2 \\
 & +\ 2.\ 5\ 0\ 5\ 8 \\
\hline
 & 4.\ 3\ 0\ 0\ 0
\end{array}
$$

Example 4

Add: 0.0357 + 0.929

Solution

$$
\begin{array}{r}
\overset{1}{}0.0357 \\
+\ 0.929 \\
\hline
0.9647
\end{array}
$$

You Try It 4

Add: 0.03294 + 0.765

Your solution

Example 5

Add: 42.3 + 162.903 + 65.0729

Solution

$$
\begin{array}{r}
\overset{111}{}42.3 \\
162.903 \\
+\ 65.0729 \\
\hline
270.2759
\end{array}
$$

You Try It 5

Add: 4.62 + 27.9 + 0.62054

Your solution

Example 6

Add: 0.83 + 7.942 + 15

Solution

$$
\begin{array}{r}
\overset{1\ 1}{}0.83 \\
7.942 \\
+\ 15. \\
\hline
23.772
\end{array}
$$

You Try It 6

Add: 6.05 + 12 + 0.374

Your solution

Example 7

Add: 9.7 thousand + 14.3 thousand

Solution

$$
\begin{array}{r}
\overset{1\ 1}{}9.7 \text{ thousand} \\
+\ 14.3 \text{ thousand} \\
\hline
24.0 \text{ thousand}
\end{array}
$$

You Try It 7

Add: 3.2 million + 5.45 million

Your solution

Solutions on p. A9

Example 8

Subtract 39.047 − 7.96 and check.

Solution

$$
\begin{array}{r}
{\overset{8}{}\overset{9}{}\overset{14}{}} \\
3\overset{8}{9}\,.\,\overset{9}{0}\,\overset{14}{4}\,7 \\
-\ 7\,.\,9\,6 \\
\hline
3\,1\,.\,0\,8\,7
\end{array}
\qquad
\text{Check:}
\qquad
\begin{array}{r}
\overset{1}{}\ \overset{1}{} \\
7.96 \\
+\ 31.087 \\
\hline
39.047
\end{array}
$$

You Try It 8

Subtract 72.039 − 8.47 and check.

Your solution

Example 9

Subtract 1.2 − 0.8235 and check.

Solution

$$
\begin{array}{r}
0\ \ 11\ 9\ 9\ 10 \\
1\,.\,2\,0\,0\,0 \\
-\ 0\,.\,8\,2\,3\,5 \\
\hline
0\,.\,3\,7\,6\,5
\end{array}
\qquad
\text{Check:}
\qquad
\begin{array}{r}
1\ 111 \\
0.8235 \\
+\ 0.3765 \\
\hline
1.2000
\end{array}
$$

You Try It 9

Subtract 3.7 − 1.9715 and check.

Your solution

Example 10

Subtract: 29 million − 9.23 million

Solution

$$
\begin{array}{r}
1\ 18\ \ \ 9\ 10 \\
2\,9\,.\,0\,0\ \text{million} \\
-\ \ 9\,.\,2\,3\ \text{million} \\
\hline
1\,9\,.\,7\,7\ \text{million}
\end{array}
$$

You Try It 10

Subtract: 35 billion − 7.67 billion

Your solution

Solutions on p. A9

Objective 3.1C   *To complete an expense form*

Employees, such as salespersons or company executives, often have expenditures for which they are reimbursed by the company. These expenditures may include meals, lodging, telephone calls, and transportation. The employee is usually required to complete an expense form, itemizing all the expenditures she or he had while on company business.

Expense forms, like many business forms, are designed to require the use of both vertical and horizontal addition so that the arithmetic involved in their completion is checked automatically. In the following expense form, the Daily Totals are found by adding the numbers in each column; the Total spent on each item is found by adding the numbers in each row. For example, the total spent on meals during the week was $84.45, and the total spent on October 3 was $263.93. The total of the Total column and the total of the Daily Total row must be the same number, because each represents the total expenditures. According to this expense form, the total expenditures from October 3 through October 10 were $670.44.

EXPENSE FORM	NAME _____ DEPARTMENT _____ LOCATION _____					
DATE	*10/3*	*10/4*	*10/5*	*10/6*	*10/7*	**TOTAL**
MEALS	*17.45*	*23.60*		*34.90*	*8.50*	*84.45*
HOTELS	*75.50*	*75.50*	*75.50*	*75.50*		*302.00*
TELEPHONE	*3.48*		*5.67*	*4.92*		*?*
ENTERTAINMENT			*83.90*			*83.90*
TRANSPORTATION	*167.50*				*5.50*	*173.00*
MISCELLANEOUS		*8.75*		*4.27*		*?*
DAILY TOTAL	*263.93*	*?*	*165.07*	*?*	*14.00*	*670.44*

Example 11

For the expense form shown above, find the total spent during the week on telephone calls and the total spent on October 4.

Strategy

* To find the total spent on telephone calls, add the numbers in the row entitled Telephone.

* To find the total spent on October 4, add the numbers in the column headed 10/4.

Solution

$3.48 + 5.67 + 4.92 = 14.07$

$23.60 + 75.50 + 8.75 = 107.85$

The total spent on telephone calls was $14.07.

The total spent on October 4 was $107.85.

You Try It 11

For the expense form shown above, find the total spent during the week on miscellaneous expenditures and the total spent on October 6.

Your strategy

Your solution

Solution on p. A9

*E*XERCISE 3.1A

Write each decimal in words.

1. 0.27   1. _____

2. 0.92   2. _____

3. 1.005   3. _____

4. 3.067   4. _____

5. 36.4   5. _____

6. 59.7   6. _____

7. 6.324   7. _____

8. 8.916   8. _____

9. 1.00001   9. _____

10. 3.0041   10. _____

Write each decimal in standard form.

11. seven hundred sixty-two thousandths   11. _____

12. two hundred ninety-five thousandths   12. _____

13. eight and three hundred four ten-thousandths   13. _____

14. four and nine hundred seven ten-thousandths   14. _____

15. three hundred four and seven hundredths   15. _____

16. eight hundred ninety-six and forty-one hundredths   16. _____

17. five and thirty-six hundredths   17. _____

18. nine and twenty-four hundredths   18. _____

19. three hundred sixty-two and forty-eight thousandths   19. _____

20. three thousand forty-eight and two thousand two ten-thousandths   20. _____

21. 5.23 million   21. _____

22. 12.4 billion   22. _____

23. 7.9 thousand   23. _____

24. 6.85 trillion   24. _____

*E*XERCISE 3.1B

Add.

25.   1.007
         + 2.1

26.   7.3
         + 9.005

27.   8.962
         + 10.599

28.   11.957
         + 9.374

29.   27.42
         + 9.765

30.   7.85
         + 29.762

31.   4.9257
           27.05
         + 9.0063

32.   8.72
           99.073
         + 2.9763

33.   62.4 + 9.827 + 692.44

34.   8 + 89.43 + 7.0659

35.   17.32 + 1.0579 + 16.5

36.   1.792 + 67 + 27.0526

37.   3.02 + 62.7 + 3.94

38.   9.06 + 4.976 + 59.6

39.   2.9 million + 6.5 million

40.   4.6 billion + 3 billion

41.   82.6 thousand + 9.9 thousand

42.   7.1 trillion + 7.2 trillion

Subtract and check.

43.   0.675
         − 0.32

44.   3
         − 1.296

45.   7.507
         − 3.419

46.   27.09
         − 7.265

47.   82.07
         − 7.354

48.   18.314
         − 9.785

49.   16.123
         − 7.457

50.   3.005
         − 1.982

25. _____

26. _____

27. _____

28. _____

29. _____

30. _____

31. _____

32. _____

33. _____

34. _____

35. _____

36. _____

37. _____

38. _____

39. _____

40. _____

41. _____

42. _____

43. _____

44. _____

45. _____

46. _____

47. _____

48. _____

49. _____

50. _____

Subtract and check.

51. 123.79 − 92.456

52. 23.4 − 0.921

53. 24.037 − 18.41

54. 26.029 − 19.31

55. 214 − 7.143

56. 16.5 − 9.7902

57. 13.2 − 8.6205

58. 92 − 19.2909

59. 7.01 − 2.325

60. 8.07 − 5.392

61. 35.7 million − 20.93 million

62. 6.2 billion − 4.95 billion

63. 19 thousand − 10.4 thousand

64. 5 trillion − 2.3 trillion

*E*XERCISE 3.1C

Complete the expense form.

EXPENSE FORM						NAME _____ DEPARTMENT _____ LOCATION _____
DATE	11/5	11/6	11/7	11/8	11/9	TOTAL
MEALS	23.45	18.25	36.90		12.75	65.
HOTELS	67.50	67.50	67.50	67.50		66.
TELEPHONE	5.42		3.64		4.96	67.
ENTERTAINMENT				78.80		68.
TRANSPORTATION	52.40				52.40	69.
MISCELLANEOUS		3.49		8.22	2.75	70.
DAILY TOTAL	71.	72.	73.	74.	75.	76.

51. _____
52. _____
53. _____
54. _____
55. _____
56. _____
57. _____
58. _____
59. _____
60. _____
61. _____
62. _____
63. _____
64. _____
65. _____
66. _____
67. _____
68. _____
69. _____
70. _____
71. _____
72. _____
73. _____
74. _____
75. _____
76. _____

Complete the expense form.

EXPENSE FORM	NAME _____ DEPARTMENT _____ LOCATION _____					
DATE	9/26	9/27	9/28	9/29	9/30	TOTAL
MEALS	27.63	19.48		31.94	8.25	77.
HOTELS	72.25	72.25	72.25	72.25		78.
TELEPHONE	4.17		8.27		6.42	79.
ENTERTAINMENT			63.40			80.
TRANSPORTATION	28.45		5.50		28.45	81.
MISCELLANEOUS		6.50		4.82		82.
DAILY TOTAL	83.	84.	85.	86.	87.	88.

Complete the expense form.

EXPENSE FORM	NAME _____ DEPARTMENT _____ LOCATION _____					
DATE	3/12	3/13	3/14	3/15	3/16	TOTAL
MEALS	17.25	23.48	19.60	27.35	31.85	89.
HOTELS	87.50	87.50	87.50	87.50	87.50	90.
TELEPHONE	5.82		4.67		6.89	91.
ENTERTAINMENT		84.90		76.25		92.
TRANSPORTATION	78.80					93.
MISCELLANEOUS		5.50	6.45			94.
DAILY TOTAL	95.	96.	97.	98.	99.	100.

77. _____

78. _____

79. _____

80. _____

81. _____

82. _____

83. _____

84. _____

85. _____

86. _____

87. _____

88. _____

89. _____

90. _____

91. _____

92. _____

93. _____

94. _____

95. _____

96. _____

97. _____

98. _____

99. _____

100. _____

SECTION 3.2 | Multiplication and Division of Decimals

Objective 3.2A   *To round a decimal to a given place value*

Rounding decimals is the same as rounding whole numbers except that the digits to the right of the given place value are dropped instead of being replaced by zeros.

If the digit to the right of the given place value is less than 5, that digit and all digits to the right are dropped. If the digit to the right of the given place value is greater than or equal to 5, increase the given place value by one and drop all digits to its right.

Round 26.3799 to the nearest hundredth.

Given place value (hundredth)

26.3799

9 is greater than 5   Increase 7 by one and drop all digits to the right of 7.

26.3799 rounded to the nearest hundredth is 26.38.

Example 1

Round 0.39275 to the nearest ten-thousandth.

Solution

Given place value (ten-thousandth)

0.39275

5 = 5

0.39275 rounded to the nearest ten-thousandth is 0.3928.

You Try It 1

Round 4.349254 to the nearest hundredth.

Your solution

Example 2

Round 42.0273 to the nearest thousandth.

Solution

Given place value (thousandth)

42.0273

3 is less than 5

42.0273 rounded to the nearest thousandth is 42.027.

You Try It 2

Round 3.2905 to the nearest thousandth.

Your solution

Solutions on p. A10

Objective 3.2B   *To multiply decimals*

Decimals are multiplied in the same way as whole numbers. Then the decimal point is placed in the product. Writing the decimals as fractions shows where to write the decimal point in the product.

$$0.3 \times 5 = \frac{3}{10} \times \frac{5}{1} = \frac{15}{10} = 1.5$$

1 decimal place   1 decimal place

$$0.3 \times 0.5 = \frac{3}{10} \times \frac{5}{10} = \frac{15}{100} = 0.15$$

1 decimal place   1 decimal place   2 decimal places

$$0.3 \times 0.05 = \frac{3}{10} \times \frac{5}{100} = \frac{15}{1000} = 0.015$$

1 decimal place   2 decimal places   3 decimal places

To multiply decimals, multiply the numbers as in whole numbers. Write the decimal point in the product so that the number of decimal places in the product is the sum of the decimal places in the factors.

Multiply: 21.4×0.36

$$
\begin{array}{rl}
21.4 & \text{1 decimal place} \\
\times\,0.36 & \text{2 decimal places} \\
\hline
1284 & \\
642 & \\
\hline
7.704 & \text{3 decimal places}
\end{array}
$$

Multiply: 0.037×0.08

$$
\begin{array}{rll}
0.037 & \text{3 decimal places} & \text{Two zeros must be inserted between the 2} \\
\times\ \ 0.08 & \text{2 decimal places} & \text{and the decimal point so that there are 5} \\
\hline
0.00296 & \text{5 decimal places} & \text{decimal places in the product.}
\end{array}
$$

Multiply 357×0.29. Round to the nearest tenth.

$$
\begin{array}{rl}
357 & \text{no decimal places} \\
\times\,0.29 & \text{2 decimal places} \\
\hline
3213 & \\
714 & \\
\hline
103.53 & \text{2 decimal places}
\end{array}
$$

$357 \times 0.29 \approx 103.5$ The symbol $\approx$ means "is approximately equal to."

To multiply by a power of 10 (10, 100, 1000, . . .), move the decimal point to the right the same number of places as there are zeros in the power of 10.

$3.8925 \times 1\underline{0}$ $= 38.925$

1 zero   move 1 decimal place

$3.8925 \times 1\underline{00}$ $= 389.25$

2 zeros   move 2 decimal places

$3.8925 \times 1\underline{000}$ $= 3892.5$

3 zeros   move 3 decimal places

$3.8925 \times 1\underline{0,000}$ $= 38,925.$

4 zeros   move 4 decimal places

$3.8925 \times 1\underline{00,000}$ $= 389,250.$ Note that a zero must be inserted before the decimal point.

5 zeros   move 5 decimal places

Example 3

Multiply: 920×3.7

Solution

$$
\begin{array}{r}
920 \\
\times\ \ 3.7 \\
\hline
644\ 0 \\
2760\ \ \\
\hline
3404.0
\end{array}
$$

You Try It 3

Multiply: 870×4.6

Your solution

Example 4

Multiply: 0.37×0.9
Round to the nearest hundredth.

Solution

$$
\begin{array}{r}
0.37 \\
\times\ \ 0.9 \\
\hline
0.333
\end{array}
$$
   $0.37 \times 0.9 \approx 0.33$

You Try It 4

Multiply: 0.28×0.7
Round to the nearest hundredth.

Your solution

Example 5

Multiply: $42.07 \times 10,000$

Solution

$42.07 \times 10,000 = 420,700$

You Try It 5

Multiply: 6.9×1000

Your solution

Solutions on p. A10

Objective 3.2C *To divide decimals*

To divide decimals, move the decimal point in the divisor to make it a whole number. Move the decimal point in the dividend the same number of places to the right. Place the decimal point in the quotient directly over the decimal point in the dividend. Then divide as in whole numbers.

Divide: $3.25\overline{)15.275}$

Step 1 $3.25.\overline{)15.27.5}$

Move the decimal point 2 places to the right in the divisor and in the dividend. Place the decimal point in the quotient.

Step 2
$$
\begin{array}{r}
4.7 \\
325.\overline{)\ 1527.5} \\
-1300 \\
\hline
227\ 5 \\
-227\ 5 \\
\hline
0
\end{array}
$$

Moving the decimal point the same number of decimal places in the divisor and dividend does not change the value of the quotient. The reason is that this process is the same as multiplying the numerator and denominator of a fraction by the same number. For the example above,

$$3.25\overline{)15.275} = \frac{15.275}{3.25} = \frac{15.275 \times 100}{3.25 \times 100} = \frac{1527.5}{325} = 325\overline{)1527.5}$$

When dividing decimals, we usually round the quotient off to a specified place value, rather than writing the quotient with a remainder. The symbol $\approx$ is used to indicate that the quotient is an approximate value after being rounded off.

Divide: $0.3\overline{)0.56}$
Round to the nearest hundredth.

$$
\begin{array}{r}
1.866 \approx 1.87 \\
0.3.\overline{)\ 0.5.600} \\
-3 \\
\hline
2\ 6 \\
-2\ 4 \\
\hline
20 \\
-18 \\
\hline
20 \\
-18
\end{array}
$$

The division must be carried to the thousandths' place in order to round the quotient to the nearest hundredth. Therefore, zeros must be inserted in the dividend so that the quotient has a digit in the thousandths' place.

Divide: $57.93 \div 3.24$
Round to the nearest thousandth.

$$
\begin{array}{r}
17.8796 \approx 17.880 \\
3.24.\overline{)\ 57.93.0000} \\
-32\ 4 \\
\hline
25\ 53 \\
-22\ 68 \\
\hline
2\ 85\ 0 \\
-2\ 59\ 2 \\
\hline
25\ 80 \\
-22\ 68 \\
\hline
3\ 120 \\
-2\ 916 \\
\hline
2040 \\
-1944
\end{array}
$$

Zeros must be inserted in the dividend so that the quotient has a digit in the ten-thousandths' place.

To divide a decimal by a power of 10 (10, 100, 1000, . . .), move the decimal point to the left the same number of places as there are zeros in the power of 10.

$$34.65 \div 1\underline{0} = 3.465$$

1 zero   move 1 decimal place

$$34.65 \div 1\underline{00} = 0.3465$$

2 zeros   move 2 decimal places

$$34.65 \div 1\underline{000} = 0.03465$$

3 zeros   move 3 decimal places

Note that a zero must be inserted between the 3 and the decimal point.

$$34.65 \div 1\underline{0,000} = 0.003465$$

4 zeros   move 4 decimal places

Note that two zeros must be inserted between the 3 and the decimal point.

Example 6

Divide: $0.1344 \div 0.032$

Solution

$$
\begin{array}{r}
4.2 \\
0.032\overline{)\ 0.134.4} \\
-\underline{128} \\
6\,4 \\
-\underline{6\,4} \\
0
\end{array}
$$

You Try It 6

Divide: $0.1404 \div 0.052$

Your solution

Example 7

Divide: $58.092 \div 82$
Round to the nearest hundredth.

Solution

$$
\begin{array}{r}
0.708 \approx 0.71 \\
82\overline{)\ 58.092} \\
-\underline{57\,4} \\
69 \\
-\underline{0} \\
692 \\
-\underline{656}
\end{array}
$$

You Try It 7

Divide: $37.042 \div 76$
Round to the nearest thousandth.

Your solution

Solutions on p. A10

Example 8

Divide: 420.9 ÷ 70.6
Round to the nearest tenth.

Solution

$$
\begin{array}{r}
5.96 \approx 6.0 \\
70.6\overline{)\,420.9{,}00} \\
-353\,0 \\
\hline
67\,90 \\
-63\,54 \\
\hline
4\,3\,60 \\
-4\,2\,36 \\
\hline
\end{array}
$$

You Try It 8

Divide: 370.2 ÷ 5.09
Round to the nearest tenth.

Your solution

Example 9

Divide 402.75 ÷ 1000

Solution

402.75 ÷ 1000 = 0.40275

You Try It 9

Divide: 42.93 ÷ 100

Your solution

Solutions on p. A10

Objective 3.2D *To convert fractions to decimals and decimals to fractions*

Every fraction can be written as a decimal. To write a fraction as a decimal, divide the numerator of the fraction by the denominator. The quotient can be rounded to the desired place value.

Convert $\frac{3}{7}$ to a decimal.

$$
\begin{array}{r}
0.42857 \\
7\overline{)3.00000}
\end{array}
$$

$\frac{3}{7}$ rounded to the nearest hundredth is 0.43.

$\frac{3}{7}$ rounded to the nearest thousandth is 0.429.

$\frac{3}{7}$ rounded to the nearest ten-thousandth is 0.4286.

Convert $3\frac{2}{9}$ to a decimal. Round to the nearest thousandth.

$$3\frac{2}{9} = \frac{29}{9} = 29 \div 9$$

$$
\begin{array}{r}
3.2222 \\
9\overline{)29.0000}
\end{array}
$$

$3\frac{2}{9}$ rounded to the nearest thousandth is 3.222.

To convert a decimal to a fraction, remove the decimal point and place the decimal part over a denominator equal to the place value of the last digit in the decimal. Write the fraction in simplest form.

hundredths

$$0.47 = \frac{47}{100}$$

hundredths

$$7.45 = 7\frac{45}{100} = 7\frac{9}{20}$$

thousandths

$$0.275 = \frac{275}{1000} = \frac{11}{40}$$

hundredths

$$0.16\frac{2}{3} = \frac{16\frac{2}{3}}{100} = 16\frac{2}{3} \div 100$$

$$= \frac{50}{3} \times \frac{1}{100} = \frac{1}{6}$$

Example 10

Convert $\frac{3}{8}$ to a decimal.

Solution

$$\begin{array}{r} 0.375 \\ 8)\overline{3.000} \end{array}$$

Example 11

Convert 0.82 to a fraction.

Solution

$$0.82 = \frac{82}{100} = \frac{41}{50}$$

You Try It 10

Convert $2\frac{3}{4}$ to a decimal.

Your solution

You Try It 11

Convert 5.35 to a fraction.

Your solution

Solutions on pp. A10–A11

*O*bjective 3.2E   *To complete a purchase order*

The operation of multiplication is used in completing a variety of business forms. Purchase orders, invoices, and sales slips are a few examples.

Purchase orders generally include the following information for each item listed: a stock number or catalog number; a description of the item; the quantity, or number ordered; the unit price; and the extension, which is the total cost for that item.

STOCK NO.	DESCRIPTION	QUANTITY	UNIT PRICE	EXTENSION
372	Computer Paper	20	28.95	579.00
408	Cartridge Ribbon	35	6.95	243.25
510	Floppy disks	100	2.50	250.00
512	Disk Holder	15	19.95	299.25
			TOTAL	$1371.50

The extension is found by multiplying the unit price of an item by the quantity. The total at the bottom of the form is the sum of the extensions.

Example 12

Calculate the extensions and the total.

QUANTITY	UNIT PRICE	EXTENSION
15	8.95	
25	12.48	
TOTAL		

You Try It 12

Calculate the extensions and the total.

QUANTITY	UNIT PRICE	EXTENSION
12	4.25	
16	9.45	
TOTAL		

Strategy

To calculate the extensions and the total:

* Multiply the unit price in the first row by the quantity in the first row.

* Multiply the unit price in the second row by the quantity in the second row.

* Add the two products.

Your strategy

Solution

```
    8.95
×     15
   44 75
   89 5
  134.25
```

```
   12.48
×     25
   62 40
  249 6
  312.00
```

```
  134.25
+ 312.00
  446.25
```

The extensions are $134.25 and $312.00.

The total is $446.25.

Your solution

Solution on p. A11

EXERCISE 3.2A

Round each decimal to the given place value.

1.	7.359	tenths		

2. 6.405   tenths

3. 89.19204   tenths

4. 480.325   hundredths

5. 670.974   hundredths

6. 22.68259   hundredths

7. 1.03925   thousandths

8. 7.072854   thousandths

9. 8.6273402   hundredths

10. 36.41859   hundredths

11. 1946.395   hundredths

12. 728.5963   hundredths

EXERCISE 3.2B

Multiply.

13. 7.7×0.9

14. 0.67×0.9

15. 2.5×5.4

16. 0.83×5.2

17. 1.47×0.09

18. 6.75×0.007

19. 0.86×0.07

20. 0.49×0.16

21. 5.41×0.7

22. 8.62×4

23. 2.19×9.2

24. 0.478×0.37

1. _____
2. _____
3. _____
4. _____
5. _____
6. _____
7. _____
8. _____
9. _____
10. _____
11. _____
12. _____
13. _____
14. _____
15. _____
16. _____
17. _____
18. _____
19. _____
20. _____
21. _____
22. _____
23. _____
24. _____

25. 0.0173×0.89 **26.** 2.437×6.1 **27.** 94.73×0.57

28. 8.005×0.067 **29.** 1.25×5.6 **30.** 89.23×0.62

31. 0.039×100 **32.** $3.57 \times 10,000$ **33.** 8.52×10

34. 6.8×1000 **35.** 64.93×100 **36.** 4.625×1000

EXERCISE 3.2C

Divide. Round to the nearest tenth.

37. $55.62 \div 8.8$ **38.** $25.43 \div 5.4$ **39.** $5.427 \div 9.5$

40. $1.837 \div 1.4$ **41.** $18.4 \div 7.3$ **42.** $52.9 \div 8.1$

43. $0.183 \div 0.17$ **44.** $0.3811 \div 0.47$ **45.** $0.542 \div 0.65$

46. $6.924 \div 0.053$ **47.** $8137 \div 1000$ **48.** $357.92 \div 10$

25. _____
26. _____
27. _____
28. _____
29. _____
30. _____
31. _____
32. _____
33. _____
34. _____
35. _____
36. _____
37. _____
38. _____
39. _____
40. _____
41. _____
42. _____
43. _____
44. _____
45. _____
46. _____
47. _____
48. _____

Divide. Round to the nearest hundredth.

49. 4.817 ÷ 16   **50.** 6.467 ÷ 8   **51.** 0.0418 ÷ 0.53

52. 0.0647 ÷ 0.72   **53.** 9 ÷ 0.48   **54.** 7 ÷ 0.55

55. 19.08 ÷ 0.45   **56.** 21.792 ÷ 0.96   **57.** 38.665 ÷ 0.95

58. 42.67 ÷ 10   **59.** 82,547 ÷ 1000   **60.** 23.627 ÷ 100

EXERCISE 3.2D

Convert the fraction to a decimal. Round to the nearest thousandth.

61. $\dfrac{5}{8}$ **62.** $\dfrac{7}{12}$ **63.** $\dfrac{2}{3}$ **64.** $\dfrac{5}{6}$

65. $\dfrac{1}{6}$ **66.** $\dfrac{7}{8}$ **67.** $\dfrac{5}{12}$ **68.** $\dfrac{9}{16}$

69. $2\dfrac{3}{1000}$ **70.** $3\dfrac{5}{10}$ **71.** $\dfrac{3}{8}$ **72.** $\dfrac{11}{16}$

Convert the decimal to a fraction.

73. 0.8   **74.** 0.4   **75.** 0.32   **76.** 0.48

49. _____
50. _____
51. _____
52. _____
53. _____
54. _____
55. _____
56. _____
57. _____
58. _____
59. _____
60. _____
61. _____
62. _____
63. _____
64. _____
65. _____
66. _____
67. _____
68. _____
69. _____
70. _____
71. _____
72. _____
73. _____
74. _____
75. _____
76. _____

77. 0.125   **78.** 0.485   **79.** 1.25   **80.** 3.75

81. 0.045   **82.** 0.085   **83.** $0.33\frac{1}{3}$ **84.** $0.66\frac{2}{3}$

EXERCISE 3.2E

Calculate the extensions and the total.

STOCK NO.	DESCRIPTION	QUANTITY	UNIT PRICE	EXTENSION
35G	Memo Pads	150	3.49	85.
43H	Cellophane tape	250	.85	86.
78M	Steno pads	75	.99	87.
92K	Ball-point pens	125	.49	88.
			TOTAL	89.

CATALOG NO.	DESCRIPTION	QUANTITY	UNIT PRICE	EXTENSION
3-432	6-pt. sockets	150	1.99	90.
3-810	8-pt. sockets	225	2.09	91.
3-755	12-pt. sockets	175	2.29	92.
9-928	Extension bars	55	2.49	93.
			TOTAL	94.

CATALOG NO.	DESCRIPTION	QUANTITY	UNIT PRICE	EXTENSION
FA-309	3.5-inch disks	100	29.99	95.
FD-355	5 1/4-inch disks	100	12.99	96.
GA-422	Disk Filer	25	14.95	97.
KB-870	Printer Buffer	10	129.95	98.
HH-265	Daisy Wheel	5	15.95	99.
			TOTAL	100.

77. _____
78. _____
79. _____
80. _____
81. _____
82. _____
83. _____
84. _____
85. _____
86. _____
87. _____
88. _____
89. _____
90. _____
91. _____
92. _____
93. _____
94. _____
95. _____
96. _____
97. _____
98. _____
99. _____
100. _____

SECTION 3.3   The Metric System of Measurement

Objective 3.3A   *To convert units of length, mass, and capacity in the metric system of measurement*

In 1789 an attempt was made to standardize units of measurement internationally in order to simplify trade and commerce between nations. A commission in France developed a system of measurement known as the **metric system.** The metric system of measurement is now used in all countries except the United States, where its use has been increasing over the past years. As shown in this section, a major benefit of the metric system is the ease of converting from one unit to another.

The basic unit of *length*, or distance, in the metric system is the **meter** (m). One meter is approximately the distance from a doorknob to the floor. All units of length in the metric system are derived from the meter. Prefixes to the basic unit denote the length of each unit. For example, the prefix "centi-" means one-hundredth; therefore, one centimeter is 1 one-hundredth of a meter.

kilo- = 1000	1 kilometer (km) = 1000 meters (m)
hecto- = 100	1 hectometer (hm) = 100 m
deca- = 10	1 decameter (dam) = 10 m
	1 meter (m) = 1 m
deci- = 0.1	1 decimeter (dm) = 0.1 m
centi- = 0.01	1 centimeter (cm) = 0.01 m
milli- = 0.001	1 millimeter (mm) = 0.001 m

Mass and weight are closely related. *Weight* is a measure of how strongly the earth is pulling on an object. Therefore, an object's weight is less in space than on the earth's surface. However, the amount of material in the object, its *mass*, remains the same. On the surface of the earth, the terms *mass* and *weight* can be used interchangeably.

The basic unit of mass in the metric system is the **gram** (g). If a box that is 1 cm long on a side is filled with water, the mass of that water is 1 gram.

1 gram = the mass of water in a box
1 cm on a side

The gram is a very small unit of mass. A paperclip weighs about one gram. The kilogram (1000 grams) is a more useful unit of mass in business applications.

The units of mass in the metric system have the same prefixes as the units of length.

1 kilogram (kg)   = 1000 grams (g)
1 hectogram (hg) = 100 g
1 decagram (dag) = 10 g
1 gram (g)   = 1 g
1 decigram (dg)   = 0.1 g
1 centigram (cg) = 0.01 g
1 milligram (mg) = 0.001 g

Liquid substances are measured in units of *capacity*. The basic unit of capacity in the metric system is the **liter** (L). One liter is defined as the capacity of a box that is 10 centimeters long on each side.

The units of capacity in the metric system have the same prefixes as the units of length.

1 kiloliter (kl)   = 1000 liters (L)
1 hectoliter (hl) = 100 L
1 decaliter (dal) = 10 L
1 liter (L)   = 1 L
1 deciliter (dl)   = 0.1 L
1 centiliter (cl) = 0.01 L
1 milliliter (ml) = 0.001 L

Converting between units in the metric system involves moving the decimal point to the right or to the left. Listing the units in order from largest to smallest will indicate how many places to move the decimal point and in which direction.

To convert 4200 cm to meters, write the units of length in order from largest to smallest.

km   hm   dam   m   dm   cm   mm   Converting cm to m requires
                                          moving 2 positions to
            2 positions   the left.

4200 cm = 42.00 m   Move the decimal point the
                                          same number of places and
            2 places   in the same direction.

Convert 324 g to kilograms.

kg   hg   dag   g   dg   cg   mg   Write the units in order from
                                          largest to smallest.

        3 positions   Converting g to kg requires
                                          moving 3 positions to the left.

324 g = 0.324 kg   Move the decimal point the
                                          same number of places and
        3 places   in the same direction.

Example 1	You Try It 1
Convert 0.38 m to millimeters.	Convert 3.07 m to centimeters.
Solution	**Your solution**
0.38 m = 380 mm	

Example 2	You Try It 2
Convert 4.23 g to milligrams.	Convert 42.3 mg to grams.
Solution	**Your solution**
4.23 g = 4230 mg	

Example 3	You Try It 3
Convert 824 ml to liters.	Convert 2 kl to liters.
Solution	**Your solution**
824 ml = 0.824 L	

Example 4	You Try It 4
Convert 4793 g to kilograms.	Convert 5682 m to kilometers.
Solution	**Your solution**
4793 g = 4.793 kg	

Solutions on p. A11

***O*bjective 3.3B**   *To calculate freight charges*

Trucks, ships, trains, aircraft, and pipelines are carriers of freight, which can include automobiles, machinery, grain, livestock, coal, building supplies, and petroleum products. Freight charges frequently are calculated on the basis of the weight of the products and the distance they are transported.

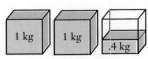

1 kg   1 kg   .4 kg

$4 per kilogram
Charged for 3 kg
$4 × 3 = $12

Freight charges may be quoted as the cost per weight unit—for example, $4 per kilogram. In that case, the weight of the shipment is rounded *up* to the nearest unit. A carrier, or transportation company, transporting 2.4 kilograms at $4 per kilogram would charge for 3 kilograms, or $12.

Example 5

Find the freight charges for transporting nine 170-liter drums of lubricating oil weighing 175.5 kg each if the charge is $26 per 100 kg.

You Try It 5

Find the freight charges for transporting 180 cartons of machine parts if each carton weighs 22.5 kg and the charge is $13 per 100 kg.

Strategy

To find the freight charges:

* Find the total weight by multiplying the number of items (9) by the weight of each item (175.5 kg).

* Divide the total weight by 100 kg, because the charge is based on the number of 100-kg "units" shipped. Round the quotient up to the nearest whole number.

* Multiply the number of 100-kg "units" transported by the charge per 100 kg ($26).

Your strategy

Solution

$$\begin{array}{r} 175.5 \\ \times \quad 9 \\ \hline 1579.5 \end{array}$$

$1579.5 \div 100 = 15.795 \approx 16$

$$\begin{array}{r} 26 \\ \times \ 16 \\ \hline 156 \\ 26 \ \\ \hline 416 \end{array}$$

The freight charges are $416.

Your solution

Solution on p. A12

EXERCISE 3.3A

Convert.

1. 42 cm = _____ mm

2. 420 g = _____ kg

3. 4200 ml = _____ L

4. 81 mm = _____ cm

5. 127 mg = _____ g

6. 3.42 L = _____ ml

7. 6804 m = _____ km

8. 4.2 kg = _____ g

9. 423 mm = _____ cm

10. 2.109 km = _____ m

11. 0.45 g = _____ mg

12. 642 m = _____ km

13. 432 cm = _____ m

14. 1856 g = _____ kg

15. 1.37 kg = _____ g

16. 42,350 g = _____ kg

17. 0.88 m = _____ cm

18. 4057 mg = _____ g

19. 0.0456 g = _____ mg

20. 2.3 kg = _____ g

21. 4.62 kl = _____ L

22. 1423 L = _____ kl

23. 2.5 km = _____ m

24. 3750 m = _____ km

25. 0.037 L = _____ ml

26. 0.035 kl = _____ L

1. _____
2. _____
3. _____
4. _____
5. _____
6. _____
7. _____
8. _____
9. _____
10. _____
11. _____
12. _____
13. _____
14. _____
15. _____
16. _____
17. _____
18. _____
19. _____
20. _____
21. _____
22. _____
23. _____
24. _____
25. _____
26. _____

Exercise 3.3B

Solve.

27. Find the freight charges for transporting 27,500 kg of corrugated aluminum sheets if the charge is $.05 per kilogram.

28. Find the freight charges for transporting 10,000 kg of potatoes if the charge is $.03 per kilogram.

29. Calculate the freight charges for transporting one million catalogs, each weighing 748 g, if the charge is $.25 per kilogram.

40. Calculate the freight charges for transporting one thousand copies of a book that weighs 862 g if the charge is $.35 per kilogram.

31. Find the freight charges for transporting 12 electric ranges, each weighing 75.5 kg, if the charge is $1.20 per kilogram.

32. Find the freight charges for transporting 25 televisions, each weighing 36.5 kg, if the charge is $.80 per kilogram.

33. A tractor trailer is transporting 8 cars to an automobile dealer. Three of the cars weigh 1258.5 kg each. The other cars weigh 1476.5 kg each. Find the total weight of the shipment of cars.

34. The Garfield Furniture Company plans to transport 5 dressers and 6 night tables by freight. Each dresser weighs 26.75 kg and each night table weighs 14.5 kg. Find the total weight of the dressers and night tables.

35. Calculate the freight charge for transporting a shipment weighing 347.8 kg if the charge is $15 per 100 kg.

36. Calculate the freight charge for transporting a shipment weighing 692.4 kg if the charge is $18 per 100 kg.

37. Find the freight charge for transporting 250 air conditioners if each air conditioner weighs 68.5 kg and the charge is $12 per 100 kg.

38. Find the freight charge for transporting 375 bags of fertilizer if each bag weighs 22.8 kg and the charge is $9 per 100 kg.

39. Calculate the freight charge for transporting 100 automobile tires, each weighing 9.1 kg, and 908 kg of auto parts if the freight charge is $7.50 per 100 kg.

40. Calculate the freight charge for transporting 50 steel girders, each weighing 136.5 kg, and 1750 kg of fabricated iron angles if the charge is $8.50 per 100 kg.

27. _____

28. _____

29. _____

30. _____

31. _____

32. _____

33. _____

34. _____

35. _____

36. _____

37. _____

38. _____

39. _____

40. _____

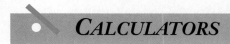

CALCULATORS

Truncated Answers   Use your calculator to multiply

$$2.43478 \times 3.2411$$

Depending on the type of calculator you are using, the answer displayed will be

   7.8913655   or   7.8913654

where the only difference is the last digit. The reason for this difference is that 2.43478 times 3.2411 is only *approximately* equal to the number in the calculator's display. The exact answer is shown below with the last three digits underlined.

$$2.43478 \times 3.2411 = 7.891365\underline{458}$$

Because the display of most calculators contains only 8 digits, the engineer who designs a calculator decides on one of two methods for displaying an answer of more than 8 digits: either to round off the answer, as in

$$7.891365458 \approx 7.8913655$$

or to *truncate* the answer, which means that the digits that cannot be displayed are discarded, as in

$$7.891365458 \approx 7.8913654$$

It is important to realize that the answer in the display of your calculator is sometimes not the exact answer to a problem and to know how your calculator handles approximate answers.

Suppose you had used your calculator to multiply 2.43478 by 3.2411. How could you determine whether the answer in the display was an approximate answer or an exact answer? (*Hint:* Consider the rules for multiplying decimals.)

How could you determine the exact answer to the multiplication shown above if your calculator displays only 8 digits? Begin by determining how many digits must be to the right of the decimal point in the answer. The number of decimal places in the product is the sum of the decimal places in the factors. There are 5 decimal places in 2.43478 and 4 decimal places in 3.2411. The number of decimal places in the product is $5 + 4 = 9$. Note that the product of the last digit in 2.4347<u>8</u> and the last digit in 3.241<u>1</u> is 8. Therefore, there will be 9 decimal places in the answer. If the product of the last digits in the factors is 0, the calculator will not display this digit in the answer.

The number displayed in the calculator (7.8913655 or 7.8913654) has 7 decimal places. The exact answer has 9 decimal places. Multiply the number in the display by 100. The result is 789.13655 or 789.13654. The decimal point was moved two places to the right. Now subtract 780 from the number in the display. This will remove the two left-most digits, thus providing space in the display for the two right-most digits. The result is 9.1365458; all the digits to the right of the decimal point are now displayed.

Use your calculator to find the exact answer.

1. 84.73214×0.6592 2. 384.2763×1.95

3. $5 \div 512$ 4. $3 \div 2560$

1. _____

2. _____

3. _____

4. _____

Sum Key   The following sales reports require two totals. The sum of the vertical totals must equal the sum of the horizontal totals. This self-checking method of completing a form is called **crossfooting** or **crosschecking**. The total of the vertical sums and the total of the horizontal sums is called the **square balance**.

Use the sum feature of your calculator to complete the forms. Consult your operator's manual for the keystroking procedures to be used on your calculator.

Name	Week 1	Week 2	Week 3	Week 4	Total
Andrews	387.10	530.69	743.86	438.22	5.
Burrows	776.76	827.68	680.51	283.46	6.
Downey	569.17	266.77	811.51	464.65	7.
Holt	615.80	489.90	763.89	816.90	8.
Manse	725.79	450.84	247.01	603.66	9.
Mitchell	498.15	368.25	831.93	219.51	10.
Wang	416.34	793.52	768.64	479.00	11.
Total	12.	13.	14.	15.	16.

Name	Mon.	Tues.	Wed.	Thurs.	Fri.	Weekly Total
Briggs	39.69	15.96	17.99	66.72	67.45	17.
Dune	54.14	88.31	28.89	14.66	28.42	18.
Farland	30.64	94.72	45.39	78.73	82.01	19.
Lange	36.96	28.41	43.66	51.32	61.90	20.
Moore	83.59	17.05	68.89	80.94	18.97	21.
Saltz	24.55	22.12	41.49	38.37	85.62	22.
Weiss	94.92	49.57	62.52	61.08	84.38	23.
Total	24.	25.	26.	27.	28.	29.

5. _____

6. _____

7. _____

8. _____

9. _____

10. _____

11. _____

12. _____

13. _____

14. _____

15. _____

16. _____

17. _____

18. _____

19. _____

20. _____

21. _____

22. _____

23. _____

24. _____

25. _____

26. _____

27. _____

28. _____

29. _____

BUSINESS CASE STUDY

Cost Estimates

You own and operate a plastics company. You produce products made to other manufacturers' specifications. You employ 10 workers; each is paid $12.50 per hour.

OCTOBER						
S	M	T	W	T	F	S
	1	2	3	4	5	6
7	8	9	10	11	12	13
14	15	16	17	18	19	20
21	22	23	24	25	26	27
28	29	30	31			

Yesterday, October 1, you received a request for an estimate from the purchasing agent of a manufacturing company. On the basis of the description of the job, you calculated that the cost for raw materials would be $4000. If you put all 10 of your employees to work on this project, the job could be completed in five 8-hour days. No overtime would be required, because the customer wants the product delivered by the fifteenth of the month. This would still leave enough time to ship the goods by motor freight. You estimate that the shipment will weigh approximately 800 kg. The freight company quoted you a price of $13.50 per 100 kg for shipment from your plant to the customer's.

Today you faxed a price estimate to the purchasing agent and within a few hours received a response. The agent's schedule has changed. Could you complete the job and have the products delivered by October 10? The agent requested a new price estimate based on the new delivery date.

You know that you cannot possibly receive the raw materials for the job before the end of the day tomorrow. If you send the shipment via air freight instead of motor freight, the shipment can be sent out at 5 o'clock on October 9 and it will still arrive by the requested delivery date. However, the shipment cost for air freight will be double that of the original estimate.

On weekdays, your employees are paid time and a half for hours worked over 8 hours a day. They are paid double time (twice the regular hourly wage) for time worked on Saturday or Sunday. The weekend falls on October 6 and 7. You would prefer not to have your employees work these days.

1. What was the original cost estimate?

2. How many hours of overtime will be required of the employees if the goods are to be shipped by 5 o'clock on October 9?

3. Is it possible to avoid having the employees work on the weekend? How might you schedule the overtime work over the days available?

4. What is the cost estimate for production and delivery of the goods by October 10?

5. What considerations should you take into account in determining whether to accept the order? Consider such issues as the present or potential relationship between your company and this customer and whether the company is in a position to stop production on any other jobs in progress. Should you consider your employees' reactions to having to work overtime on such short notice?

CHAPTER SUMMARY

Key Words
The number 26.45 is in **decimal notation**. The 26 is the **whole-number part**, the period is the **decimal point**, and the 45 is the **decimal part** of the number. (Objective 3.1A)

The symbol ≈ means "is approximately equal to." (Objective 3.2C)

In the **metric system** of measurement, the **meter** is the basic unit of length, the **gram** is the basic unit of mass, and the **liter** is the basic unit of capacity. (Objective 3.3A)

Essential Rules
To write a decimal in words, write the decimal part as though it were a whole number. Then name the place value of the last digit. (Objective 3.1A)

To write a decimal in standard form, insert zeros as necessary after the decimal point so that the last digit is in the given place-value position. (Objective 3.1A)

To add or subtract decimals, write the numbers so that the decimal points are lined up vertically. Add or subtract as for whole numbers, and write the decimal point in the answer directly below the decimal points in the problem. (Objective 3.1B)

To round a decimal, round as for whole numbers. Then drop the digits to the right of the given place value instead of replacing them by zeros. (Objective 3.2A)

To multiply decimals, multiply the numbers as in whole numbers. Write the decimal point in the product so that the number of decimal places in the product is the sum of the numbers of decimal places in the factors. (Objective 3.2B)

To multiply by a power of 10, move the decimal point to the right the same number of places as there are zeros in the power of 10. (Objective 3.2B)

To divide decimals, move the decimal point in the divisor to make it a whole number. Move the decimal point in the dividend the same number of places to the right. Place the decimal point in the quotient directly over the decimal point in the dividend. Then divide as in whole numbers. (Objective 3.2C)

To divide a decimal by a power of 10, move the decimal point to the left the same number of places as there are zeros in the power of 10. (Objective 3.2C)

To convert a fraction to a decimal, divide the numerator of the fraction by the denominator. (Objective 3.2D)

To convert a decimal to a fraction, remove the decimal point and place the decimal part over a denominator equal to the place value of the last digit in the decimal. (Objective 3.2D)

To find the extension on a purchase order, multiply the unit price of an item by the quantity ordered. (Objective 3.2E)

Prefixes in the metric system (Objective 3.3A)

kilo-	= 1000	deci-	= 0.1
hecto-	= 100	centi-	= 0.01
deca-	= 10	milli-	= 0.001

To convert between units in the metric system, list the units in order from largest to smallest. Move the decimal point in the given number the same number of places and in the same direction as is required to move from the given unit to the unit to which you are converting. (Objective 3.3A)

REVIEW / TEST

1. Write 45.0302 in words.

2. Write two hundred nine and seven thousand eighty-six hundred-thousandths in standard form.

3. Write 11.8 million in standard form.

4. Add:
270.93 + 97 + 1.976 + 88.675

5. Subtract: 37.003 − 9.23674

6. Add: 10.2 million + 3.75 million

7. Round 0.07395 to the nearest thousandth.

8. Multiply: 0.369×6.7

9. Multiply: 58.9×1000

10. Divide: $0.0659 \div 0.037$
Round to the nearest thousandth.

11. Divide: $61,924 \div 10,000$

12. Convert $\frac{4}{5}$ to a decimal.

13. Convert 0.875 to a fraction.

14. Convert 6.4 to a fraction.

15. Convert 730 g to kilograms.

16. Convert 820 cm to meters.

1. _____

    _____

    _____

    _____

2. _____

3. _____

4. _____

5. _____

6. _____

7. _____

8. _____

9. _____

10. _____

11. _____

12. _____

13. _____

14. _____

15. _____

16. _____

17. Find the freight charges for transporting 250 packages of insulating material if each package weighs 16.5 kg and the charge is $6.50 per 100 kilograms.

Complete the expense form.

EXPENSE FORM			NAME _____			
			DEPARTMENT _____			
			LOCATION _____			

DATE	12/2	12/3	12/4	12/5	12/6	TOTAL
MEALS	12.78	9.42			10.56	18.
HOTELS						19.
TELEPHONE		3.47		4.63		20.
ENTERTAINMENT						21.
TRANSPORTATION		4.75			3.50	22.
MISCELLANEOUS	2.85			3.16		23.
DAILY TOTAL	24.	25.	26.	27.	28.	29.

Calculate the extensions and the total of the purchase order.

ITEM NO.	DESCRIPTION	QUANTITY	UNIT PRICE	EXTENSION
2X-113	Battery Charger	15	29.99	30.
2X-125	Gas shocks	25	9.99	31.
3Z-247	Engine Analyzer	10	49.99	32.
			TOTAL	33.

17. _____

18. _____

19. _____

20. _____

21. _____

22. _____

23. _____

24. _____

25. _____

26. _____

27. _____

28. _____

29. _____

30. _____

31. _____

32. _____

33. _____

4

Banking

*F*OR MOST PEOPLE, the first bank account they open is a checking account. A checking account safeguards your money, helps you resist the temptation posed by a large amount of cash in your pocket, and makes bill paying easier. Using a checking account wisely also helps you to build a good relationship with your bank, which is important if you ever decide to apply for a loan.

Most banks offer two types of accounts: regular checking accounts and NOW accounts. A regular checking account does not pay interest on your balance. (Interest is the subject of Chapter 9.) Generally, there is a monthly maintenance charge, a charge per check, or both. However, if you maintain a specified minimum balance, these charges may not apply. For those who are opening an account with a small deposit, a regular checking account is usually the best option.

NOW accounts pay interest on the balance in the account. A minimum balance of $500 or more may be required. As long as the balance meets this minimum, no check or service charges are paid on the account.

Many banks offer other services in conjunction with a checking account. Through **automatic savings**, you specify an amount of money to be transferred from your checking account to your savings account at whatever interval you choose. Through **automatic drafting**, you can have bills such as a monthly car payment or mortgage payment paid automatically each month. Through **direct deposit**, funds such as a paycheck are sent directly to the bank for deposit, thus saving time for the account holder and guaranteeing that the money is in the account quickly. An account holder with **overdraft protection** has an agreement with the bank that if the account is short of the funds needed to make payment on a check, the necessary funds will be borrowed.

Because the terms and conditions of checking accounts differ, it is wise to "check out" different banks and get answers to a number of questions.

1. What are the fees associated with a checking account? Examples include charges for writing checks, maintaining the account, bouncing a check, stopping payment on a check, using an automated teller machine (ATM) card, and transferring funds by telephone.
2. What banking services are available through using the ATM card, what machines accept it, and where are they located?
3. What other services are available with the checking account?

**SECTION 4.1**   **Checking Accounts**

*O*bjective 4.1A   *To write checks and deposit slips*

A checking account can be opened at most banks by depositing an amount of money in the bank. When this first deposit is made, the depositor fills out a **signature card.** The signature card is kept in the bank's files and provides the bank with a copy of the official signatures of those persons authorized to sign checks.

Title	East Phoenix Rental Equipment	Account Number

In consideration of the acceptance by Meyers' National Bank of my/our account of the type indicated below. I/we agree to be bound by such rules and regulations and/or such schedules of interest, fees and charges applicable to such account as may now or hereafter be adopted by and in effect at said Bank, and also by the provisions printed hereon. It is under-stood that the acceptance by said Bank of my/our account is subject to the receipt by said Bank of satisfactory credit information.

(1) Sign Here   *Eugene L. Madison*

(2) Sign Here   *Gloria B. Masters*

Address   3011 N.W. Ventura Street

City   Phoenix   State   Arizona   Zip   85280

☑ CHECKING   ☐ MULTIPLE MATURITY   ☐ CASH MANAGER
☐ SAVINGS   ☐ GUARANTEED INTEREST ☐ SAFE DEPOSIT   ☐ OTHER _____
                 (Multiple Maturity)

IF THIS IS A JOINT ACCOUNT, BOTH OWNERS MUST SIGN ABOVE

Each of the signers guarantees the genuineness of the signature of the other. Each signer also agrees with the other and the Bank that deposits now or hereafter made to this account may be withdrawn in whole or part by either or survivor, and that each may endorse for deposit to this account any instrument payable to the order of either or both. Provisions respecting this agreement shall be modified only upon receipt by the Bank of written notice, signed by both.

The depositor receives from the bank a checkbook containing checks and deposit slips. Checks are used to direct the bank to pay money to a designated person or company, called the **payee.** Upon paying the check, the bank deducts the amount of the check from the depositor's account.

DATE CHECK IS WRITTEN   **CHECK NUMBER**

East Phoenix Rental Equipment   NO. 2023
3011 N.W. Ventura Street   68 - 461   ← **ABA**
Phoenix, Arizona 85280   **PAYEE**   1052   **NUMBER**

                                        October 11 , 19 *94*

PAY TO THE
ORDER OF   *Tellas Manufacturing Co.* $ *827 00/100*   ← **AMOUNT OF CHECK**

Eight Hundred Twenty-Seven and 00/100 ~~~~~ DOLLARS

MEYERS' NATIONAL BANK
11 N.W. Nova Street
Phoenix, Arizona 85215

                                                        DEPOSITOR'S SIGNATURE

                            Eugene L. Madison

Memo _____
I: 1052 III 0461 I: 5008 2023 II'

AMOUNT OF CHECK IN WORDS

For a check to be deposited or to be cashed, it must be endorsed on the back left end of the check. A **restrictive endorsement** is used for checks to be deposited. "For Deposit Only" is written on the back of the check, followed by the payee's account number, and the signature of the payee. This endorsement restricts the use of the check; it can be deposited only in the payee's account.

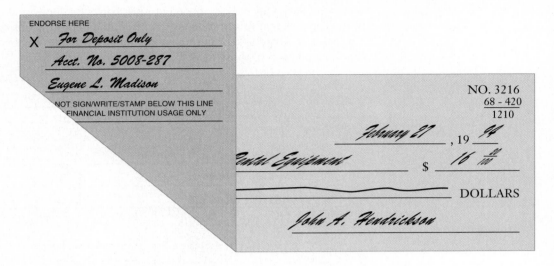

The payee for the check shown above is East Phoenix Rental Equipment. Note on the previous page that Eugene L. Madison and Gloria B. Masters signed the signature card for the East Phoenix Rental Equipment checking account. Therefore, either Madison's or Master's signature is a valid endorsement.

A **blank endorsement** is used for checks to be cashed; only the payee's signature is written on the back of the check. The check then becomes payable to anyone who presents it; therefore, it is as good as cash. A blank endorsement should only be used when the account holder is actually at the bank.

In a **full** or **special endorsement**, the payee writes the words "Pay to the Order of" on the back of the check, followed by the name of the individual or company to whom the check is to be payable, and the payee's signature. Only the individual or company named in the endorsement can then cash or deposit the check. The person or organization named may then use the check as though they were the original payee.

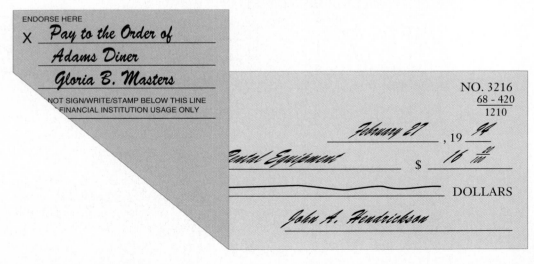

The **deposit slips** provided by the bank are used to record currency, coin, and checks to be deposited in the account. Each check listed for deposit should be identified by its American Bankers Association transit number (ABA number). The ABA number is printed as a fraction in the upper right-hand corner of a check. Only the numerator of the ABA number need be listed on the deposit slip.

	East Phoenix Rental Equipment						

CURRENCY		117	00
COIN		14	65
C H E C K S	LIST SINGLY		
	68-420	16	20
	66-164	37	29
	68-837	146	82
	62-111	107	96
	66-722	98	37
TOTAL		538	29

DEPOSIT TICKET

DATE *March 1*, 19 *94*

MEYERS' NATIONAL BANK
11 N.W. Nova Street
Phoenix, Arizona 85215

BE SURE
EACH ITEM
IS PROPERLY
ENDORSED

I:1052 ⑴0461 I:5008 287 ⑴▪

ITEMS CREDITED SUBJECT TO VERIFICATION AND DEPOSIT AGREEMENT OF THIS BANK

Besides checks and deposit slips, a new checking account holder may receive an **automated teller machine (ATM)** card and be given a **personal identification number (PIN).** The ATM card can be inserted into a slot at an automated teller machine, or terminal, and the PIN punched in at the keyboard located there. The account holder can then make certain bank transactions, such as obtaining cash or making a deposit. Customers have access to the ATM machine 24 hours a day.

An ATM is only one method of **electronic funds transfer (EFT).** An EFT system is used to perform financial transactions through a computer terminal or telephone hookup. For example, **point-of-sale (POS) terminals** are located at business establishments such as grocery stores and gasoline stations. POS terminals work in much the same way as an ATM machine: the customer inserts the card and punches in the PIN. The amount the customer has rung up at the cash register is entered into the machine, and, after the computer has verified that there is enough money in the customer's account, the dollar amount of the purchase is immediately transferred from the customer's account to the seller's account.

Personal checking accounts are offered to individuals. **Business checking accounts** are available to businesses. A business checking account operates in the same manner as a personal checking account. However, a business account may receive more services than a personal checking account. For example, a bank may arrange to receive payments on debts owed to the business, automatically crediting the amount to the business checking account.

The types of financial institutions that offer checking accounts include banks, savings and loan associations, mutual savings banks, and credit unions. A **commercial bank** is a corporation, whose primary goal, like that of other businesses, is to earn a profit.

Savings and loan associations (S & Ls) have the primary purpose of accepting savings and providing home loans. However, since 1981 they have also been able to offer their customers NOW accounts. **NOW** stands for **negotiable order of withdrawal.** A NOW account is a checking account that can earn interest on the amount deposited.

Mutual savings banks are similar to savings and loan associations; they offer consumer loans and NOW accounts to customers. However, a mutual savings bank is owned by its depositors, and the profits are distributed to them.

A **credit union** has the specific function of serving its members, all of whom have some common bond, such as the same employer or union. Credit unions are not-for-profit organizations that accept deposits and provide loans for consumer products. The credit union version of a check is called a **share draft**. A share draft looks just like a bank check and operates in the same manner.

Example 1

Write a check dated September 28, 1994, to Acton Carpet Cleaners in the amount of $187.50. Sign the check for Gloria B. Masters.

Solution

East Phoenix Rental Equipment 3011 N.W. Ventura Street Phoenix, Arizona 85280	NO. 2025 68- 461 1052

September 28, 19 *94*

PAY TO THE
ORDER OF   *Acton Carpet Cleaners* $ *187 50/100*

One Hundred Eighty-Seven and 50/100 ———————— DOLLARS

Meyers' National Bank
11 N.W. Nova Street
Phoenix, Arizona 85215

Gloria B. Masters

Memo ————————————
I:1052 III 0461 I:5008 2025 II•

You Try It 1

Write a check dated September 30, 1994, to the First National Bank in the amount of $694.80. Sign the check for Eugene L. Madison.

Your solution

East Phoenix Rental Equipment
3011 N.W. Ventura Street
Phoenix, Arizona 85280

NO. 2029
68 - 461
1052

————————— , 19 ——

PAY TO THE
ORDER OF ———————————————— $ ——————

———————————————————————— DOLLARS

Meyers' National Bank
11 N.W. Nova Street
Phoenix, Arizona 85215

Memo ————————————
I:1052 III 0461 I:5008 2029 II•

Solution on p. A13

Example 2

You are depositing, for the East Phoenix Rental Company, $125 in currency and two checks, one for $217.50 with an ABA number of 62-111, and the other for $86.99 with an ABA number of 66-722. The date is September 29, 1994. Complete the deposit slip.

Solution

East Phoenix Rental Equipment	CURRENCY	125	00
	COIN		
	C LIST SINGLY 62-111	217	50
	H 66-722	86	99
DATE *September 29*, 19 94	E		
	C		
MEYERS' NATIONAL BANK	K		
11 N.W. Nova Street	S		
Phoenix, Arizona 85215	TOTAL	429	49

DEPOSIT TICKET

BE SURE EACH ITEM IS PROPERLY ENDORSED

I:1052 '''0461 I:5008 287 ''•

ITEMS CREDITED SUBJECT TO VERIFICATION AND DEPOSIT AGREEMENT OF THIS BANK

You Try It 2

You are depositing, for the East Phoenix Rental Company, $220 in currency and two checks, one for $335.00 with an ABA number of 68-837, and the other for $74.28 with an ABA number of 66-164. The date is October 4, 1994. Complete the deposit slip.

Your solution

East Phoenix Rental Equipment	CURRENCY		
	COIN		
	C LIST SINGLY		
	H		
DATE _____ , 19 ____	E		
	C		
MEYERS' NATIONAL BANK	K		
11 N.W. Nova Street	S		
Phoenix, Arizona 85215	TOTAL		

DEPOSIT TICKET

BE SURE EACH ITEM IS PROPERLY ENDORSED

I:1052 '''0461 I:5008 287 ''•

ITEMS CREDITED SUBJECT TO VERIFICATION AND DEPOSIT AGREEMENT OF THIS BANK

Solution on p. A13

Objective 4.1B   *To calculate a checkbook balance*

In some checkbooks, each check in the depositor's checkbook is attached to a **check stub,** which is used to record the check number, the amount, the date, the payee, and the purpose of the check. The balance brought forward, deposits made since the last check was written, and the balance carried forward are also recorded on the check stub. The **balance brought forward** is the amount of money in the account prior to any transactions recorded on the check stub. The **balance carried forward** is the amount of money in the account after the transactions recorded on the check stub.

NO. _2023_ $ _827.00_

DATE _October 11_ , 19 _94_

TO _Tellas Mfg. Co._

FOR _Cultivator_

	DOLLARS	CENTS
BAL. BRO'T FOR'D	6952	95
AMT. DEPOSITED		
AMT. DEPOSITED		
TOTAL		
AMT. THIS CHECK	827	00
BAL. CAR'D FOR'D	6125	95

East Phoenix Rental Equipment
3011 N.W. Ventura Street
Phoenix, Arizona 85280

NO. 2023

68 - 461
―――――
1052

October 11 , 19 *94*

PAY TO THE
ORDER OF _Tellas Manufacturing Co._ $ _827 00/100_

Eight Hundred Twenty-Seven and 00/100 ―――― DOLLARS

Meyers' National Bank
11 N.W. Nova Street
Phoenix, Arizona 85215

Eugene L. Madison

Memo ―――――――――――

I: 1052 ʺʹ 0461 I: 5008 2023 ʺʹ

Some banks provide depositors with a **check register** rather than with check stubs. A portion of a checkbook register is shown below. The account holder had a checkbook balance of $587.93 before writing two checks, one for $286.87 and the other for $102.38, and making a deposit of $345.00. The current checkbook balance is calculated by subtracting the amount of each check from the previous balance and adding the amount of the deposit.

RECORD ALL CHARGES OR CREDITS THAT AFFECT YOUR ACCOUNT

CHECK NUMBER	DATE	DESCRIPTION OF TRANSACTION	AMOUNT OF CHECK (−)	✓ T	FEE (IF ANY) (−)	AMOUNT OF DEPOSIT (+)	BALANCE
							$ 587 93
1307	8/4	Greenwood Co.	$ 286 87		$	$	286 87
							301 06
1308	8/6	Barton Co., Inc.	102 38				102 38
							198 68
	8/6	Deposit				345 00	345 00
							543

A checkbook register is also used to record deposits or withdrawals made at an ATM machine. For transactions made at a POS terminal, the account holder records the date, where the expenditure was made, and the amount. Then the checkbook balance is calculated.

Example 3	You Try It 3
Jacob Zucker had a checking account balance of $602.46 before withdrawing $150 from the account at an ATM machine and making two deposits, one in the amount of $176.86 and the other in the amount of $94.73. Find Jacob's current checkbook balance.	Kara Tanamachi had a checking account balance of $785.93 before making a $189.43 purchase with an ATM card, writing a check for $352.68, and making a deposit of $250. Find Kara's current checkbook balance.

Strategy

To find the checkbook balance:

◆ Subtract the amount of the withdrawal from the previous balance.

◆ Add the amount of each deposit.

Your strategy

Solution

```
  602.46
 −150.00
  452.46
  176.86
 + 94.73
  724.05
```

The checkbook balance is $724.05.

Your solution

Solution on p. A14

Example 4

Calculate the balance carried forward for the check stub shown below.

NO. 1437		$ *268.00*
DATE *5/12* , 19 *94*		
TO *Harrison Co.*		
FOR *Software*		

	DOLLARS	CENTS
BAL. BRO'T FOR'D	*893*	*47*
AMT. DEPOSITED	*350*	*42*
AMT. DEPOSITED		
TOTAL		
AMT. THIS CHECK		
BAL. CAR'D FOR'D		

Strategy

To calculate the balance carried forward:

* Add the deposit to the balance brought forward.
* Subtract the amount of the check written.

Solution

```
   893.47
+ 350.42
  1243.89
-  268.00
   975.89
```

The balance carried forward is $975.89.

You Try It 4

Calculate the balance carried forward for the check stub shown below.

NO. 962		$ *146.50*
DATE *11/14* , 19 *94*		
TO *Wilson Brothers*		
FOR *Utilities*		

	DOLLARS	CENTS
BAL. BRO'T FOR'D	*773*	*28*
AMT. DEPOSITED	*294*	*63*
AMT. DEPOSITED		
TOTAL		
AMT. THIS CHECK		
BAL. CAR'D FOR'D		

Your strategy

Your solution

Solution on p. A14

EXERCISE 4.1A

1. Write a check dated January 24, 1995, to Xerox Corporation in the amount of $145.90. Sign the check for Gloria B. Masters.

East Phoenix Rental Equipment
3011 N.W. Ventura Street
Phoenix, Arizona 85280

NO. 2847

$\frac{68 - 461}{1052}$

_____ , 19 ____

PAY TO THE
ORDER OF _____ $ _____

_____ DOLLARS

Meyers' National Bank
11 N.W. Nova Street
Phoenix, Arizona 85215

Memo _____

I: 1052 ⑴ 0461 I: 5008 2847 ⑴•

2. You are depositing, for the East Phoenix Rental Company, $275 in currency and two checks, one for $398.75 with an ABA number of 62-420, and the other for $63.50 with an ABA number of 53-7159. The date is January 25, 1995. Complete the deposit slip.

East Phoenix
Rental Equipment

DEPOSIT TICKET

DATE _____ , 19 ____

MEYERS' NATIONAL BANK
11 N.W. Nova Street
Phoenix, Arizona 85215

I: 1052 ⑴ 0461 I: 5008 287 ⑴•

CURRENCY		
COIN		
C H E C K S — LIST SINGLY		
TOTAL		

BE SURE
EACH ITEM
IS PROPERLY
ENDORSED

ITEMS CREDITED SUBJECT TO VERIFICATION AND DEPOSIT AGREEMENT OF THIS BANK

EXERCISE 4.1B

3. Calculate the balance carried forward for the check stub shown below.

4. Calculate the balance carried forward for the check stub shown below.

3. _____

4. _____

NO. _____ $ *193.87*
DATE _____ , 19 ____
TO _____
FOR _____

	DOLLARS	CENTS
BAL. BRO'T FOR'D	1126	20
AMT. DEPOSITED	250	00
AMT. DEPOSITED		
TOTAL		
AMT. THIS CHECK		
BAL. CAR'D FOR'D		

NO. _____ $ *327.46*
DATE _____ , 19 ____
TO _____
FOR _____

	DOLLARS	CENTS
BAL. BRO'T FOR'D	2189	52
AMT. DEPOSITED	351	06
AMT. DEPOSITED	195	27
TOTAL		
AMT. THIS CHECK		
BAL. CAR'D FOR'D		

Complete the balance column in the check register.

RECORD ALL CHARGES OR CREDITS THAT AFFECT YOUR ACCOUNT								
CHECK NUMBER	DATE	DESCRIPTION OF TRANSACTION	AMOUNT OF CHECK (–)	✓ T	FEE (IF ANY) (–)	AMOUNT OF DEPOSIT (+)	BALANCE $892 46	
918	7/16	Insurance	$ 247 63		$	$		5.
								6.
919	7/18	Rent	550 00					7.
								8.
	7/18	Deposit				678 49		9.
								10.

Complete the balance column in the check register.

RECORD ALL CHARGES OR CREDITS THAT AFFECT YOUR ACCOUNT								
CHECK NUMBER	DATE	DESCRIPTION OF TRANSACTION	AMOUNT OF CHECK (–)	✓ T	FEE (IF ANY) (–)	AMOUNT OF DEPOSIT (+)	BALANCE $1247 63	
1284	4/25	Telephone	$ 289 57		$	$		11.
								12.
	4/25	Deposit				461 06		13.
								14.
1285	4/27	Electric Bill	43 92					15.
								16.
	4/27	ATM withdrawal	75 00					17.
								18.

5. _____
6. _____
7. _____
8. _____
9. _____
10. _____
11. _____
12. _____
13. _____
14. _____
15. _____
16. _____
17. _____
18. _____
19. _____
20. _____

Solve.

19. Don Glover had a checking account balance of $3476.85 before withdrawing $250 from the account at an ATM machine, writing a check for $848.37, and making a deposit of $1048.53. Find Don's current checkbook balance.

20. Michele Gabrielle's checking account had a balance of $1894.32 before she made a purchase of $187.46 with an ATM card and made two deposits, one in the amount of $162.42 and the other in the amount of $259.83. Find the current balance in Michele's checking account.

Bank Statements

Objective 4.2A   *To reconcile a bank statement*

Each month the bank sends a bank statement to the account holder. The bank statement lists the bank balance from the previous month's statement, the checks that the bank has paid, the deposits it has received, any transactions made at an ATM or POS terminal, and the current bank balance on the account.

@ First National Bank

First National Bank
P.O. Box 1243
Philadelphia, PA 19104

David M. Lance
P.O. Box 5647
Philadelphia, PA 19104   11/09/94   through   12/08/94

===
PERSONAL CHECKING ACCOUNT 1846579
===

DESCRIPTION	DEBITS	CREDITS	DATE	BALANCE
OPENING BALANCE			11/09/94	918.93
DEPOSIT		64.32	11/13/94	983.25
CHECK #124	112.02		11/14/94	871.23
CHECK #123	65.00		11/16/94	806.23
ATM WITHDRAWAL #003973	40.00		11/16/94	766.23
CHECK #120	116.50		11/20/94	649.73
CHECK #121	41.53		11/21/94	608.20
ATM WITHDRAWAL #009723	70.00		11/23/94	538.20
CHECK #130	86.29		11/24/94	451.91
CHECK #128	25.18		11/25/94	426.73
DEPOSIT		80.00	11/25/94	506.73
CHECK #122	109.94		11/27/94	396.79
ATM WITHDRAWAL #000498	40.00		11/30/94	356.79
ATM WITHDRAWAL #004766	60.00		12/03/94	296.79
CHECK #125	23.98		12/05/94	272.81
ATM WITHDRAWAL #002998	30.00		12/06/94	242.81
SERVICE FEE	6.50			236.31
CLOSING BALANCE				236.31

To **reconcile** a bank statement means to determine whether the checking account balance is accurate. Reconciliation requires a number of steps.

1. List the current checkbook balance.
2. Add to the current checkbook balance all checks that have been written but have not yet been paid by the bank. These checks are called **checks outstanding.** Also add any additions to the balance noted on the bank statement. For example, some banks pay account holders interest on the money deposited in a bank account.
3. Subtract any deposits that have not yet been recorded by the bank. These deposits are called **deposits in transit.** Also subtract any charges noted on the bank statement, such as service charges for check processing and record keeping. This is the checkbook balance.
4. Compare the balance with the bank balance listed on the bank statement. If the two numbers are equal, the bank statement and the checkbook balance.

Example 1

Find the current bank balance.

Current Checkbook Balance

$1493.76

Checks Outstanding

$243.87
$ 98.42
$165.99

Deposits in Transit

$647.28

Solution

$1493.76
243.87
98.42
+   165.99
$2002.04
−   647.28
$1354.76

The current bank balance is $1354.76.

You Try It 1

Find the current bank balance.

Current Checkbook Balance

$2157.93

Checks Outstanding

$324.66
$ 49.81
$155.79

Deposits in Transit

$838.45

Your solution

Solution on p. A14

Example 2

Reconcile the bank statement.

STATEMENT OF ACCOUNT	**MEYERS' NATIONAL BANK**	

East Phoenix Rental Equipment
3011 N.W. Ventura Street
Phoenix, Arizona 85280

ACCOUNT NO.   *5008-287*
STATEMENT DATE   *October 31,1994*

BALANCE BROUGHT FORWARD FROM LAST STATEMENT	6495.13
NEW BALANCE	8393.81

DATE	CHECKS, WITHDRAWALS, PAYMENTS	DEPOSITS	BALANCE
10/1	50.00	921.00	7366.13
10/4	200.00		7166.13
10/6	400.00	463.62	7229.75
10/11	174.23		7055.52
10/15	61.68	789.43	7783.27
10/19	591.84	1063.14	8254.57
10/23	817.22		7437.35
10/25	250.00	1211.96	8399.31
10/31	5.50 Service Charge		8393.81

NO.	DATE	TRANSACTION	BALANCE
			6495.13
2143	9/27	Cash	50.00
			6445.13
	9/29	Deposit	921.00
			7366.13
	9/30	ATM Withdrawal	200.00
			7166.13
2144	10/3	Rent	400.00
			6766.13
	10/5	Deposit	463.62
			7229.75
2145	10/9	Newport Co.	174.23
			7055.52
2146	10/12	Monroe Corp.	61.68
			6993.84
	10/14	Deposit	789.43
			7783.27

NO.	DATE	TRANSACTION	BALANCE
			7783.27
2147	10/16	Richmond Co.	591.84
			7191.43
	10/18	Deposit	1063.14
			8254.57
2148	10/21	Jasper Services	817.22
			7437.35
2149	10/22	Lincoln Press	189.50
			7247.85
	10/23	ATM Withdrawal	250.00
			6997.85
	10/24	Deposit	1211.96
			8209.81
2150	10/29	Telephone	347.62
			7862.19
2151	10/31	Deposit	274.83
			8137.02

Solution

1. List the current checkbook balance: $8137.02

2. Add all checks outstanding and any additions to the balance: Check No. 2149   189.50
                                                                    Check No. 2150   347.62
                                                                    Subtotal: 8674.14

3. Subtract any deposits in transit and any bank charges: 274.83
                                                                    Subtotal: 8399.31

4. The balance should agree with the balance listed on the   5.50
   bank statement.   Balance: $8393.81

You Try It 2

Reconcile the bank statement.

STATEMENT OF ACCOUNT	MEYERS' NATIONAL BANK		

East Phoenix Rental Equipment
3011 N.W. Ventura Street
Phoenix, Arizona 85280

ACCOUNT NO.	5008-287	
STATEMENT DATE	November 30,1994	

BALANCE BROUGHT FORWARD FROM LAST STATEMENT	8393.81	
NEW BALANCE	8838.90	

DATE	CHECKS, WITHDRAWALS, PAYMENTS	DEPOSITS	BALANCE
11/1	189.50	247.83	8452.14
11/3	347.62		8104.52
11/5	400.00	684.51	8389.03
11/10	209.71		8179.32
11/16	50.00	508.27	8637.59
11/20	539.62	927.16	9025.13
11/24	343.08	162.35	8844.40
11/30	5.50 Service Charge		8838.90

NO.	DATE	TRANSACTION	BALANCE
			8393.81
1763	10/28	Lincoln Press	189.50
			8204.31
	10/31	Deposit	247.83
			8452.14
1764	11/1	Richmond Co.	347.62
			8104.52
1765	11/3	Rent	400.00
			7704.52
	11/4	Deposit	684.51
			8389.03
1766	11/7	Newport Co.	209.71
			8179.32
	11/14	ATM Withdrawal	50.00
			8129.32
	11/15	Deposit	508.27
			8637.59

NO.	DATE	TRANSACTION	BALANCE
			8637.59
1767	11/17	Monroe Corp.	539.62
			8097.97
	11/19	Deposit	927.16
			9025.13
1768	11/21	Telephone	343.08
			8682.05
	11/23	Deposit	162.35
			8844.40
1769	11/27	Lincoln Press	216.84
			8627.56
1770	11/29	Jasper Services	938.36
			7689.20
	11/30	Deposit	416.59
			8105.79

Your solution

1. List the current checkbook balance:
2. Add all checks outstanding and any additions to the balance:

3. Subtract any deposits in transit and any bank charges:

4. The balance should agree with the balance listed on the bank statement.

Check No. ___ _____
Check No. ___ _____
Subtotal: _____

Subtotal: _____

Balance: _____

Solution on p. A15

EXERCISE 4.2A

Find the current bank balance.

1. | Current Checkbook Balance $387.42 | Checks Outstanding $47.60 | Deposits in Transit $128.97 | Service Charge $1.75 |

2. | Current Checkbook Balance $465.91 | Checks Outstanding $131.85 | Deposits in Transit $276.50 | Service Charge $3.25 |

3. | Current Checkbook Balance $609.84 | Checks Outstanding $225.39 | Deposits in Transit $311.47 | Interest Earned $2.28 |

4. | Current Checkbook Balance $1357.84 | Checks Outstanding $114.63 $ 84.97 $286.59 | Deposits in Transit $526.92 | Interest Earned $5.11 |

5. | Current Checkbook Balance $2943.24 | Checks Outstanding $387.45 $476.02 $ 92.18 | Deposits in Transit $642.93 | Service Charge $5.50 |

6. | Current Checkbook Balance $976.83 | Checks Outstanding $ 47.92 $502.66 $275.83 | Deposits in Transit $284.27 $319.60 | Service Charge $3.00 |

7. | Current Checkbook Balance $3854.32 | Checks Outstanding $ 347.91 $1087.62 $ 567.53 | Deposits in Transit $1243.86 $ 319.60 | Interest Earned $14.46 |

8. | Current Checkbook Balance $2068.41 | Checks Outstanding $378.29 $550.00 $264.06 $143.75 | Deposits in Transit $575.50 $380.95 | Service Charge $2.50 |

1. _____
2. _____
3. _____
4. _____
5. _____
6. _____
7. _____
8. _____

Reconcile the bank statement.

STATEMENT OF ACCOUNT	FIRST NATIONAL BANK		
Lennox Company	**ACCOUNT NO.** *6093-167* **STATEMENT DATE** *June 30, 1994*		

BALANCE BROUGHT FORWARD FROM LAST STATEMENT	2873.66
NEW BALANCE	2493.42

DATE	CHECKS, WITHDRAWALS, PAYMENTS	DEPOSITS	BALANCE
6/2	550.00	129.32	2452.98
6/5	147.81		2305.17
6/7	250.00	387.55	2442.72
6/10	108.64		2334.08
6/17	50.00		2284.08
6/21	357.09	542.19	2469.18
6/23	218.64	247.63	2498.17
6/27	4.75 Service Charge		2493.42

NO.	DATE	TRANSACTION	BALANCE
			2873.66
892	6/1	Rent	550.00
			2323.66
	6/2	Deposit	129.32
			2452.98
893	6/3	Advertising	147.81
			2305.17
894	6/5	Bryant Rental	250.00
			2055.17
	6/7	Deposit	387.55
			2442.72
895	6/8	Utilities	108.64
			2334.08
	6/16	ATM Withdrawal	50.00
			2284.08
896	6/18	Telephone	357.09
			1926.99

NO.	DATE	TRANSACTION	BALANCE
			1926.99
	6/20	Deposit	542.19
			2469.18
897	6/20	Insurance	218.64
			2250.54
898	6/22	Newton Co.	386.25
			1864.29
	6/22	Deposit	247.63
			2111.92
899	6/27	C.R. Restaurant	192.14
			1919.78
	6/29	Deposit	408.25
			2328.03

1. List the current checkbook balance:
2. Add all checks outstanding and any additions to the balance:

3. Subtract any deposits in transit and any bank charges:

4. The balance should agree with the balance listed on the bank statement.

Check No. _____
Check No. _____
Subtotal: _____

Subtotal: _____

Balance: _____

9. _____
10. _____
11. _____
12. _____
13. _____
14. _____
15. _____
16. _____

CALCULATORS

The Digit 0   **When the following numbers are used in a calculation, which digits, if any, need not be entered on the calculator?**

1. 1.50
2. 1.05
3. 0.51
4. 0.50
5. 10.05
6. 10.00
7. 15.00

1. _____
2. _____
3. _____
4. _____
5. _____
6. _____
7. _____

BUSINESS CASE STUDY

Choosing a Bank

You own and operate a travel agency. Until now, you have had only one office, and you have run the business on your own. You are the travel agent for a number of firms in the area, and that portion of your business is fairly steady. However, the portion of your business that caters to vacationers is seasonal; people tend to vacation in the spring, in the summer, and around the New Year's holiday. Consequently, the balance in your checking account can fluctuate quite a bit. Because your balance can fall below any given minimum at certain times of the year, you chose a regular account for your business checking account.

On your business account, you currently pay a $3 monthly service fee and a cost of $.10 per check. Each month you write checks for the rent on the office space, the telephone, the electricity, the car payment on your business vehicle, and your newspaper advertisements. You have two credit cards, used mostly to charge office supplies; the monthly bill for each of these you pay by check. Every three months you receive bills for your auto insurance and your health insurance; you pay each of these by check as well. The annual expenses that you pay by check include car registration, dues to the professional organization to which you belong, a subscription to a trade journal, and the fee charged by the accountant who prepares your tax forms. All other expenses, such as the cost of gasoline, you pay in cash. You have never had an overdraft. (An **overdraft** occurs when a check is written for which there are insufficient funds in the checking account.)

You recently decided that it would be profitable to open a second office in the next city. Because you do not have the capital to start up the operation, you went to the bank at which you have your business checking account to apply for a business loan, but the loan officer told you that you do not qualify for a loan. You decided to apply for a loan at another bank, and your loan was approved. You learn that this bank offers an account that has no monthly service charge during a month in which your balance is over $500. If your balance falls below $500 during a given month, the service fee is $5. There is a cost of $.20 per check for each check written in excess of 10 checks per month. The bank offers automatic drafting, through which you can have your office rental payment, your car payment, and payments on your business loan made automatically. The bank will also automatically transfer funds from the business account to your personal checking account each month. With this account, you would have an ATM card. There is no charge for use of the ATM card at the bank's terminal; the fee for using the ATM card at any other machine is $.50.

1. What is your present business checking account costing you per year? Assuming that your balance will fall below $500 only four months per year and that you will generally incur an ATM charge twice a month, find the costs per year that you would incur if you were to switch your account to the other bank. What is the difference in cost between the two banks?

2. It has been stated that the five criteria to apply in choosing a bank are cost, convenience, safety, treatment of customers, and range of services. Both banks are members of the FDIC. (The **Federal Deposit Insurance Corporation** insures all accounts in a member bank for up to $100,000 per depositor.) The bank that approved your loan is five miles farther from your office than the bank at which you presently have an account. Taking into account all of these criteria, rate each of the banks.

CHAPTER SUMMARY

Key Words   A **signature card** is filled out by anyone opening a checking account. The signature card is kept in the bank's files in order to provide the bank with a copy of the official signatures of those persons authorized to sign checks. (Objective 4.1A)

Checks are used to direct the bank to pay money to a designated person or company, called the **payee**. Upon paying the check, the bank deducts the amount of the check from the depositor's account. **Deposit slips** are used to record currency, coin, and checks to be deposited in the account. (Objective 4.1A)

Sometimes a **check stub** is attached to each check. A check stub is used to record the check number, the amount, the date, the payee, and the purpose of the check. The balance brought forward, deposits made since the last check was written, and the balance carried forward are also recorded on the check stub. Instead of a check stub, depositors may be provided with a **check register.** (Objective 4.1B)

A **bank statement** is sent to the account holder each month. The bank statement lists the bank balance from the previous month's statement, the checks that the bank has paid, the deposits it has received, any transactions made at an ATM or POS terminal, and the current bank balance on the account. (Objective 4.2A)

To **reconcile a bank statement** means to determine whether the checking account balance is correct. (Objective 4.2A)

Checks outstanding are checks that have been written by the account holder but have not yet been paid by the bank. **Deposits in transit** are deposits that have not yet been recorded by the bank. (Objective 4.2A)

Essential Rules   On a **deposit slip**, each check listed should be identified by its ABA number. Only the numerator of the ABA number is written. (Objective 4.1A)

To calculate the current checkbook balance, add any deposits to the previous balance, and subtract from the balance any checks written and withdrawals or purchases made with an ATM card. (Objective 4.1B)

To reconcile a bank statement:

1. List the current checkbook balance.
2. Add to the current checkbook balance the checks outstanding and any additions to the balance noted on the bank statement.
3. Subtract any deposits in transit and any charges noted on the bank statement.
4. Compare the balance with the bank balance listed on the bank statement. If the two numbers are equal, the checkbook and the bank statement balance. (Objective 4.2A)

REVIEW / TEST

1. Write a check dated February 8, 1995, to White River Company in the amount of $342.97. Sign the check for Gloria B. Masters.

East Phoenix Rental Equipment
3011 N.W. Ventura Street
Phoenix, Arizona 85280

NO. 2936

$\dfrac{68 - 461}{1052}$

_____ , 19 ____

PAY TO THE
ORDER OF _____ $ _____

_____ DOLLARS

Meyers' National Bank
11 N.W. Nova Street
Phoenix, Arizona 85215

Memo _____   _____

I: 1052 ⑴ 0461 I: 5008 2936 ⑴

---

2. You are depositing, for the East Phoenix Rental Company, $150 in currency and two checks, one for $247.50 with an ABA number of 62-111, and the other for $199.98 with an ABA number of 66-722. The date is February 10, 1995. Complete the deposit slip.

DEPOSIT TICKET

East Phoenix
Rental Equipment

DATE _____ , 19 _____

MEYERS' NATIONAL BANK
11 N.W. Nova Street
Phoenix, Arizona 85212

I: 1052 ⑴ 0461 I: 5008 287 ⑴

CURRENCY		
COIN		
CHECKS (LIST SINGLY)		
TOTAL		

BE SURE EACH ITEM IS PROPERLY ENDORSED

ITEMS CREDITED SUBJECT TO VERIFICATION AND DEPOSIT AGREEMENT OF THIS BANK

---

3. Calculate the balance carried forward for the check stub shown below.

4. Calculate the balance carried forward for the check stub shown below.

3. _____

4. _____

NO. _____ $ *319.62*	DOLLARS	CENTS
DATE _____ , 19 ___		
TO _____		
FOR _____		
BAL. BRO'T FOR'D	*847*	*25*
AMT. DEPOSITED	*268*	*47*
AMT. DEPOSITED		
TOTAL		
AMT. THIS CHECK		
BAL. CAR'D FOR'D		

NO. _____ $ *493.27*	DOLLARS	CENTS
DATE _____ , 19 ___		
TO _____		
FOR _____		
BAL. BRO'T FOR'D	*1680*	*55*
AMT. DEPOSITED	*297*	*48*
AMT. DEPOSITED	*162*	*09*
TOTAL		
AMT. THIS CHECK		
BAL. CAR'D FOR'D		

5. Joyce Bannen had a checking account balance of $1267.84 before making a deposit of $392.50, making a $124.33 purchase with an ATM card, and writing a check for $872.15. Find Joyce's current checkbook balance.

6. José Ocasio's checking account had a balance of $618.43 before he made a purchase of $93.48 with an ATM card and made two deposits, one in the amount of $277.90 and the other in the amount of $175. Find the current balance of José's checking account.

5. _____

6. _____

7. _____

8. _____

9. _____

10. _____

11. _____

12. _____

13. _____

14. _____

15. _____

16. _____

Complete the balance column in the check register.

CHECK NUMBER	DATE	DESCRIPTION OF TRANSACTION	AMOUNT OF CHECK (−)	✓ T	FEE (IF ANY) (−)	AMOUNT OF DEPOSIT (+)	BALANCE
RECORD ALL CHARGES OR CREDITS THAT AFFECT YOUR ACCOUNT							$ *947* 62
	9/6	*Deposit*	$		$	$ *351* 08	7.
							8.
647	9/9	*Kingston Co.*	593 47				9.
							10.
	9/11	*Deposit*				260 89	11.
							12.
	9/15	*ATM Withdrawal*	150 00				13.
							14.

Find the current bank balance.

15.
Current Checkbook Balance	Checks Outstanding	Deposits in Transit	Service Charge
$1286.43	$239.70 $ 91.87 $306.45	$481.93 $162.58	$4.00

16.
Current Checkbook Balance	Checks Outstanding	Deposits in Transit	Interest Earned
$1945.36	$432.38 $647.92 $ 35.25 $183.44	$549.27 $392.64	$7.29

5

Equations

Objectives

5.1A   To determine whether a given number is a solution
of an equation

5.1B   To solve an equation

5.2A   To solve a problem using a formula

5.3A   To express a quantity in terms of a variable

5.3B   To solve application problems

*Y*OU ENCOUNTER PROBLEM-SOLVING situations every day. Some problems you may mentally solve without considering the steps you are taking. For example, suppose a friend suggests you both take a trip over spring break. You'd like to go. What questions go through your mind? You might ask yourself the following:

How much will the trip cost (for travel, hotel rooms, meals, and so on)?

Are my friend and I going to share some costs?

Can I afford it?

How much money do I have in the bank?

How much more money than I have now do I need for the trip?

How much time do I have to earn that money?

How much can I earn in that amount of time?

How much money must I keep in the bank in order to pay the next tuition bill?

These questions require different mathematical skills. Determining the cost of the trip requires *estimation;* you must use your knowledge of air fares or the cost of gasoline to arrive at an estimate of these costs. If some of the costs are going to be shared, you need to *divide* those costs by 2 in order to determine your share of the expense. Determining how much more money you need requires *subtraction:* the amount needed minus the amount currently in the bank. To determine how much money you can earn in the given amount of time requires *multiplication* of the amount you earn per week by the number of weeks you can work.

Facing this problem-solving situation may not seem difficult to you, because you have faced similar situations before and therefore know what questions to ask.

An important aspect of learning to solve problems is learning what questions to ask. As you work through the application problems in this chapter, try to become more conscious of the mental process you are engaging in. You might begin the process by asking:

1. Have I read the problem enough times to understand the situation being described?
2. Will restating the problem in my own words help me to understand the problem situation better?
3. What facts are given? (Make a list of the information given in the problem.)
4. What information is being asked for?
5. What is the relationship between the given facts and the solution?

Try to focus on the problem-solving situation, not on getting the answer quickly. And remember, the more problems you solve, the better able you become to solve other problems, partly because you are learning what questions to ask.

128

SECTION 5.1 Solving Equations

*O*bjective 5.1A *To determine whether a given number is a solution of an equation*

Often we discuss a quantity without knowing its exact value—for example, the price of gold next month or the market price of a share of stock next week. When a letter of the alphabet is used to stand for a quantity that is unknown, the letter is called a **variable**.

An **equation** states that two expressions are equal. Here is an example of an equation.

$x + 4 = 9$ In this equation, x is a variable.
The equation states that a number plus 4 equals 9.

Although some equations are more complicated than this one, this equation does have the main characteristics shared by all equations. There is a *left side* $(x + 4)$, an *equals sign* (=), and a *right side* (9).

Here is another example of an equation:

$2y = 6$ In this equation, y is a variable.
The expression $2y$ means "2 times y" or "$2 \cdot y$."
The equation states that 2 times a number equals 6.

Just as a sentence may be true or false, an equation may be true or false.

The equation at the right is true if
the variable is replaced by 5.

$x + 4 = 9$
$5 + 4 = 9$ True equation

The equation is false if the variable
is replaced by 2.

$x + 4 = 9$
$2 + 4 = 9$ False equation

A **solution** of an equation is a number that, when substituted for the variable, results in a true equation.

5 is a solution of the equation $x + 4 = 9$ because $5 + 4 = 9$ is a true equation.
2 is not a solution of the equation $x + 4 = 9$ because $2 + 4 = 9$ is a false equation.

To determine whether a given number is a solution of an equation, replace the variable in the equation with the given number. Then simplify. If the result is a true equation, the given number is a solution of that equation.

Is 0.8 a solution of the equation $1.5t = 1.2$?

◆ Replace the variable t with 0.8.

◆ Simplify 1.5 times 0.8.

◆ Compare the results. If the results are equal,
the given number is a solution of the equation.
If the results are not equal, the given number
is not a solution of the equation.

$$\frac{1.5t}{1.5\,(0.8)} \Big|\, \frac{1.2}{1.2}$$
$$1.2 = 1.2$$

Yes, 0.8 is a
solution of the
equation $1.5t = 1.2$.

Example 1

Is 3.4 a solution of the equation
$8.7 + r = 12.6$?

Solution

$$\begin{array}{c|c} 8.7 + r & = 12.6 \\ \hline 8.7 + 3.4 & 12.6 \\ 12.1 & \neq 12.6 \end{array}$$ The symbol $\neq$ means "is not equal to."

No, 3.4 is not a solution of the
equation $8.7 + r = 12.6$.

You Try It 1

Is 10.4 a solution of the equation
$4.6 = D - 5.8$?

Your solution

Example 2

Is 21 a solution of the equation $\frac{2}{3} M = 14$?

Solution

$$\begin{array}{c|c} \frac{2}{3} M & = 14 \\ \hline \frac{2}{3}(21) & 14 \\ 14 & = 14 \end{array}$$

Yes, 21 is a solution of the equation

$\frac{2}{3} M = 14$.

You Try It 2

Is $\frac{4}{5}$ a solution of the equation $10p = 6$?

Your solution

Solutions on p. A16

*O*bjective 5.1B   *To solve an equation*

To **solve an equation** means to find the solution of the equation.

Solve the equation $P = 2 + 5$.

♦ The variable P is alone on the left side of
the equation.
Simplify the right side of the equation.

$P = 2 + 5$
$P = 7$

♦ When $P = 7$, $P = 2 + 5$ is a true equation.

The solution is 7.

The solution of the equation
$x + 3 = 8$ is 5 because when
5 is substituted for x, the
result is a true equation.

$$x + 3 = 8$$
$$5 + 3 = 8$$
$$8 = 8$$

Note that if 4 is added to each
side of the equation, the
solution is still 5.

$$x + 3 = 8$$
$$x + 3 + 4 = 8 + 4$$
$$x + 7 = 12 \qquad 5 + 7 = 12$$

If 2 is subtracted from each
side of the equation, the
solution is still 5.

$$x + 3 = 8$$
$$x + 3 - 2 = 8 - 2$$
$$x + 1 = 6 \qquad 5 + 1 = 6$$

This illustrates two important properties of equations.

> **The same number can be added to each side of an equation without changing the solution of the equation.**

> **The same number can be subtracted from each side of an equation without changing the solution of the equation.**

These properties are used in solving equations.

Note the effect of subtracting 3
from each side of the equation
$x + 3 = 8$.
A number plus 0 equals the
number, so $x + 0 = x$.

$$x + 3 = 8$$
$$x + 3 - 3 = 8 - 3$$
$$x + 0 = 5$$
$$x = 5$$

The value of the variable equals 5, the solution of the equation.

In solving an equation, the goal is to get the variable alone on one side of the equation; the other side of the equation is the solution. For the example above, the variable x is alone on the left side of the equation. The right side (5) is the solution.

To solve an equation in which a number is added to a variable, subtract that number from each side of the equation.

Solve the equation $S + 10 = 12$.

⬥ 10 is added to the variable S.
Subtract 10 from each side of
the equation.

$$S + 10 = 12$$
$$S + 10 - 10 = 12 - 10$$
$$S + 0 = 2$$

⬥ S is alone on the left side. The
right side (2) is the solution.

$$S = 2$$

⬥ Check the solution.

The solution checks.

$$\frac{S + 10 = 12}{2 + 10 \mid 12}$$
$$12 = 12$$

The solution is 2.

Solve: $13.8 = 5.6 + T$

♦ 5.6 is added to the variable *T*.
Subtract 5.6 from each side of
the equation.

$$13.8 = 5.6 + T$$
$$13.8 - 5.6 = 5.6 - 5.6 + T$$
$$8.2 = T$$

♦ Check the solution.

$$\frac{13.8 = 5.6 + T}{13.8 \mid 5.6 + 8.2}$$

The solution checks.

$$13.8 = 13.8$$

The solution is 8.2.

To solve an equation in which a number is subtracted from a variable, add that number to each side of the equation.

Solve: $F - 4 = 23$

♦ 4 is subtracted from the variable *F*.
Add 4 to each side of the equation.

$$F - 4 = 23$$
$$F - 4 + 4 = 23 + 4$$
$$F = 27$$

♦ Check the solution.

$$\frac{F - 4 = 23}{27 - 4 \mid 23}$$

The solution checks.

$$23 = 23$$

The solution is 27.

Solve: $2.4 = L - 8.1$

♦ 8.1 is subtracted from the variable *L*.
Add 8.1 to each side of the equation.

$$2.4 = L - 8.1$$
$$2.4 + 8.1 = L - 8.1 + 8.1$$
$$10.5 = L$$

♦ Check the solution.

$$\frac{2.4 = L - 8.1}{2.4 \mid 10.5 - 8.1}$$

The solution checks.

$$2.4 = 2.4$$

The solution is 10.5.

In each of these examples, a number was either added to or subtracted from a variable. Now we'll consider equations in which a variable is multiplied by a number.

The solution of the equation
$12x = 24$ is 2 because when 2
is substituted for *x*, the
result is a true equation.

$$12x = 24$$
$$12 \cdot 2 = 24$$
$$24 = 24$$

If we multiply each side of
the equation by 3, the solution
is still 2.

$$12x = 24$$
$$3(12x) = 3(24)$$
$$36x = 72 \qquad\qquad 36 \cdot 2 = 72$$

If we divide each side of the equation by 3, the solution is still 2.	$12x = 24$ $\dfrac{12x}{3} = \dfrac{24}{3}$

$$4x = 8 \qquad\qquad 4 \cdot 2 = 8$$

If we divide each side of the equation by 4, the solution is still 2.	$12x = 24$ $\dfrac{12x}{4} = \dfrac{24}{4}$

$$3x = 6 \qquad\qquad 3 \cdot 2 = 6$$

This illustrates two more important properties of equations.

> **Each side of an equation can be multiplied by the same number (except 0) without changing the solution of the equation.**

> **Each side of an equation can be divided by the same number (except 0) without changing the solution of the equation.**

These properties are used in solving equations.

Note the effect of dividing each side of the equation $12x = 24$ by 12.	$12x = 24$ $\dfrac{12x}{12} = \dfrac{24}{12}$
A number multiplied by 1 equals the number, so $1 \cdot x = x$.	$1x = 2$ $x = 2$

The variable equals 2, the solution of the equation.

To solve an equation in which a variable is multiplied by a number, divide each side of the equation by the number.

Solve: $0.7R = 35$

♦ *R* is multiplied by 0.7.

$$0.7R = 35$$

Divide each side of the equation by 0.7.

$$\frac{0.7R}{0.7} = \frac{35}{0.7}$$

$$\frac{0.7R}{0.7} = \frac{0.7}{0.7}R = 1R = R$$

$$\frac{35}{0.7} = 35 \div 0.7 = 50 \qquad\qquad R = 50$$

♦ Check the solution.

$$\frac{0.7R = 35}{0.7(50) \mid 35}$$
$$35 = 35$$

The solution checks.

The solution is 50.

Solve: $11 = 0.25c$

♦ c is multiplied by 0.25.

Divide each side of the equation by 0.25.

$$11 = 0.25c$$
$$\frac{11}{0.25} = \frac{0.25c}{0.25}$$
$$44 = c$$

♦ Check the solution.

The solution checks.

$$11 = 0.25c$$
$$11 \mid 0.25(44)$$
$$11 = 11$$

The solution is 44.

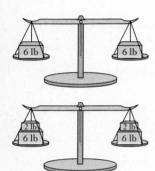

Keep in mind that in an equation, the expression on the left side is equal to the expression on the right side. In order to keep the expressions equal, whatever we do to one side of an equation we must also do to the other side.

Each side of an equation represents a number. $6 = 6$

If we add a number to one side, we must add the same number to the other side in order to keep the expressions equal. $6 + 2 = 6 + 2$

If we divide one side by a number, we must divide the other side by the same number in order to keep the expressions equal. $\dfrac{6}{3} = \dfrac{6}{3}$

Example 3

Solve: $3.62 + n = 8.17$

Solution

$$3.62 + n = 8.17$$
$$3.62 - 3.62 + n = 8.17 - 3.62$$
$$n = 4.55$$

The solution is 4.55.

Example 4

Solve: $86 = B(0.2)$

Solution

$$86 = B(0.2)$$
$$\frac{86}{0.2} = \frac{B(0.2)}{0.2}$$
$$430 = B$$

The solution is 430.

You Try It 3

Solve: $5.73 = A - 9.46$

Your solution

You Try It 4

Solve: $W(0.25) = 75$

Your solution

Solutions on p. A16

EXERCISE 5.1A

Solve.

1. Is 5.5 a solution of $x + 2.3 = 7.8$?

2. Is 9.6 a solution of $12.2 - y = 2.6$?

3. Is 2.4 a solution of $3n = 7.4$?

4. Is 6.1 a solution of $15.5 = 2.5V$?

5. Is 20 a solution of $15 = \frac{3}{4}T$?

6. Is $\frac{2}{5}$ a solution of $25P = 10$?

EXERCISE 5.1B

Solve.

7. $d + 5 = 7$

8. $y + 3 = 9$

9. $b - 4 = 11$

10. $z - 6 = 10$

11. $2 + A = 8$

12. $5 + L = 12$

13. $10 = 4 + c$

14. $12 = 3 + w$

15. $5 = m - 3$

16. $7 = n - 2$

17. $5R = 15$

18. $4T = 28$

19. $20 = 4D$

20. $18 = 2S$

1. _____
2. _____
3. _____
4. _____
5. _____
6. _____
7. _____
8. _____
9. _____
10. _____
11. _____
12. _____
13. _____
14. _____
15. _____
16. _____
17. _____
18. _____
19. _____
20. _____

21. $N = 3.9 + 8.72$

22. $M = 261.5 - 77.9$

23. $a = 49.6 - 25.7$

24. $f = 0.342 + 0.164$

25. $F + 1.7 = 5.4$

26. $V + 3.6 = 9.2$

27. $p - 8.1 = 4.3$

28. $r - 2.7 = 3.6$

29. $\dfrac{5}{8} = s + \dfrac{1}{8}$

30. $\dfrac{5}{6} = x + \dfrac{1}{6}$

31. $6.4t = 128$

32. $2.7v = 3.24$

33. $54 = 0.15W$

34. $36 = 0.75C$

35. $3486 = 2910 + P$

36. $12{,}475 = 8602 + B$

37. $62.58 = G - 47.30$

38. $902.74 = H - 381.66$

39. $88 = d(0.55)$

40. $75 = b(1.2)$

41. $A(3.6) = 90$

42. $L(4.2) = 63$

43. $874.35 = R + 362.19$

44. $620.47 = T + 493.55$

21. _____

22. _____

23. _____

24. _____

25. _____

26. _____

27. _____

28. _____

29. _____

30. _____

31. _____

32. _____

33. _____

34. _____

35. _____

36. _____

37. _____

38. _____

39. _____

40. _____

41. _____

42. _____

43. _____

44. _____

●SECTION 5.2   Formulas

Objective 5.2A   *To solve a problem using a formula*

Formulas are frequently used in business situations. In a business formula, variables are used to represent unknown quantities such as cost and profit.

To solve a problem that involves use of a formula, substitute the known quantities into the given formula. Then solve the resulting equation. Remember that the goal is to get the variable alone on one side of the equation.

The value of an investment after one year was $500. The original investment was $450. Find the increase in value of the investment by using the formula $A = P + I$, where A is the value of the investment after one year, P is the original investment, and I is the increase in value of the investment.

◆ Replace the variables A and P in the formula by their given values. ($A = 500$, $P = 450$)

$$A = P + I$$
$$500 = 450 + I$$

◆ Solve the equation for the variable I by subtracting 450 from each side of the equation.

$$500 - 450 = 450 - 450 + I$$
$$50 = I$$

The increase in value of the investment is $50.

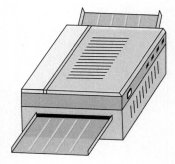

If it costs you $.04 to operate a 400-watt copy machine for 1 hour, what is your cost per kilowatt-hour? Use the formula $c = \frac{1}{1000}wtk$, where c is the cost of operating an appliance, w is the number of watts, t is the time in hours, and k is the cost per kilowatt-hour.

◆ Replace the variables c, w, and t in formula by their given values. ($c = 0.04$, $w = 400$, $t = 1$)

$$c = \frac{1}{1000}wtk$$
$$0.04 = \frac{1}{1000}(400)(1)k$$

◆ Multiply $\frac{1}{1000}$, 400, and 1 on the righthand side. Write the product as a decimal. ($\frac{400}{1000} = 400 \div 1000$)

$$0.04 = 0.4k$$

◆ Solve the equation for the variable k by dividing each side of the equation by 0.4.

$$\frac{0.04}{0.4} = \frac{0.4k}{0.4}$$
$$0.1 = k$$

◆ Write the answer in dollars.

$$k = \$.10$$

The cost per kilowatt-hour is $.10.

Example 1

A total of $900 is to be repaid in 6 equal monthly payments. Find the monthly payment by using the formula $A = P \cdot N$, where A is the amount to be repaid, P is the monthly payment, and N is the number of months.

Strategy

To find the monthly payment:

* Replace the variables A and N in the formula by their given values. ($A = 900$, $N = 6$)

* Solve the equation for P.

Solution

$$A = P \cdot N$$
$$900 = P \cdot 6$$
$$\frac{900}{6} = \frac{P \cdot 6}{6}$$
$$150 = P$$

The monthly payment is $150.

You Try It 1

A total of $600 is to be repaid in 4 equal monthly payments. Find the monthly payment by using the formula $A = P \cdot N$, where A is the amount to be repaid, P is the monthly payment, and N is the number of months.

Your strategy

Your solution

Example 2

Walden's Pharmacy has contracted to purchase 350 lines of newspaper advertising at a cost of $297.50. What is the rate charged per line? Use the formula $C = N \cdot R$, where C is the advertising cost, N is the length of the advertisement in number of lines, and R is the rate per line.

Strategy

To find the rate per line:

* Replace the variables C and N in the formula by their given values. ($C = 297.50$, $N = 350$)

* Solve the equation for R.

Solution

$$C = N \cdot R$$
$$297.50 = 350 \cdot R$$
$$\frac{297.50}{350} = \frac{350 \cdot R}{350}$$
$$0.85 = R$$

The rate per line is $.85.

You Try It 2

West Valley Jewelers has contracted to purchase 260 lines of newspaper advertising at a cost of $213.20. What is the rate charged per line? Use the formula $C = N \cdot R$, where C is the advertising cost, N is the length of the advertisement in number of lines, and R is the rate per line.

Your strategy

Your solution

Solutions on p. A17

EXERCISE 5.2A

Solve. Use the formula $A = P \cdot N$, where A is the amount to be repaid, P is the monthly payment, and N is the number of months.

1. You must pay a total of $864 in 12 equal monthly payments. Find your monthly payment.

2. You must pay a total of $3000 in 8 equal monthly payments. Find your monthly payment.

3. Maribeth Bakke must pay a total of $5400 in monthly payments of $180. How many monthly payments will Maribeth make on the debt?

4. Gene Connery must pay a total of $4500 in monthly payments of $375. In how many months will the $4500 debt be paid off?

1. _____

2. _____

3. _____

4. _____

5. _____

6. _____

7. _____

8. _____

9. _____

10. _____

Solve. Use the formula $A = P + I$, where A is the value of the investment after one year, P is the original investment, and I is the increase in value of the investment.

5. The value of a stock investment you hold is $3000. The original value of the investment one year ago was $1800. Find the increase in the value of your investment.

6. The value of a stock investment you hold is $5700. The original value of the investment one year ago was $4900. Find the increase in the value of your investment.

7. The value of an investment after one year was $17,000. The increase in value during the year was $3500. Find the amount of the original investment.

8. The value of an investment after one year was $23,000. The increase in value during the year was $2500. Find the amount of the original investment.

Solve. Use the formula $E = V - L$, where E is the equity, V is the value of the home, and L is the loan amount on the property.

9. Amanda Chicopee owns a home valued at $125,000. She has $67,853.25 in loans on the property. What is the equity in Amanda's home?

10. Juan Samora owns a home valued at $240,000. He has $142,976.80 in loans on the property. What is the equity in Juan's home?

Solve. Use the formula $E = V - L$, where E is the equity, V is the value of the home, and L is the loan amount on the property.

11.   The equity value of Louis DiSalvo's home is $28,500 and the value of the property is $167,000. What is the loan amount on Louis's home?

12.   The equity value of Irene Braun's home is $36,500 and the value of the property is $98,000. What is the loan amount on Irene's home?

Solve. Use the formula $M = S - D$, where M is the mortgage loan amount, S is the selling price, and D is the down payment.

13.   Cindy Romero made a down payment of $14,500 on a home. She took out a mortgage loan on the house for $130,500. What was the selling price of the home?

14.   John Jaworski made a down payment of $23,600 on a home. He took out a mortgage loan on the house for $94,400. What was the selling price of the home?

Solve. Use the formula $C = N \cdot R$, where C is the advertising cost, N is the length of the advertisement in number of lines, and R is the rate per line.

15.   Southshore Realty purchased 480 lines of newspaper advertising at a cost of $340.80. What was the rate charged per line?

16.   The White Mountain Ski Lodge has contracted to purchase 140 lines of newspaper advertising at a cost of $95.20. What is the rate charged per line?

Solve. Use the formula $c = \frac{1}{1000} wtk$, where c is the cost of operating an appliance, w is the number of watts, t is the time in hours, and k is the cost per kilowatt-hour.

17.   What would it cost you to operate an 1800-watt TV set for 5 hours at a cost of $.06 per kilowatt-hour?

18.   What would it cost you to operate a 200-watt stereo for 3 hours at a cost of $.10 per kilowatt-hour?

19.   If it costs you $.08 to operate a 250-watt CD player for 4 hours, what is your cost per kilowatt-hour?

20.   If it costs you $.28 to operate a 2000-watt air conditioner for 2 hours, what is your cost per kilowatt-hour?

11. _____

12. _____

13. _____

14. _____

15. _____

16. _____

17. _____

18. _____

19. _____

20. _____

SECTION 5.3　Application Problems

Objective 5.3A　*To express a quantity in terms of a variable*

Many business applications require that you identify the unknown quantity, assign a variable to that quantity, and then attempt to express other unknowns in terms of that quantity.

A confectioner makes a mixture of candy that contains 3 pounds more of milk chocolate than of caramel. Express the amount of milk chocolate in the mixture in terms of the amount of caramel in the mixture.

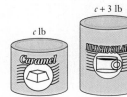

◆ Assign a variable to the amount of caramel in the mixture.	Amount of caramel in the mixture: c
◆ Express the amount of milk chocolate in the mixture in terms of c.	Amount of milk chocolate in the mixture: $c + 3$

In this example, any variable could be used for the amount of caramel in the mixture. For example, A might be chosen to represent the amount of caramel in the mixture; then the amount of milk chocolate in the mixture is $A + 3$.

Example 1

A "double-clocked" computer operates at twice its normal speed. Express the speed of a "double-clocked" computer in terms of its normal speed.

Strategy

To express the speed of a "double-clocked" computer in terms of its normal speed:

◆ Assign a variable to the normal speed of the computer.

◆ Express the "double-clocked" speed in terms of the assigned variable.

Solution

Normal speed of the computer: s

The "double-clocked" speed is twice the normal speed: $2s$

You Try It 1

The sale price of a suit is three-fourths of the regular price. Express the sale price in terms of the regular price.

Your strategy

Your solution

Solution on p. A17

*O*bjective 5.3B *To solve application problems*

An equation states that two expressions are equal. Therefore, to translate a word problem into an equation requires recognition of the words or phrases that mean "equals." Some of these words are listed below. Each translates to "=."

equals	**is**
amounts to	**represents**

A new laser printer prints at a rate of 16 ppm (pages per minute). This is four times the rate of a slower laser printer. Find the rate of the slower printer.

* Assign a variable to the rate of the slower printer.

* Write two expressions for the same value.

* Write and solve an equation.

rate of the slower printer: r

$\boxed{16}$ is $\boxed{\text{four times the rate of the slower printer}}$

$$16 = 4r$$
$$\frac{16}{4} = \frac{4r}{4}$$
$$4 = r$$

The rate of the slower printer is 4 ppm.

Example 2

An office building was purchased today for $250,000. This is 2.5 times its purchase price five years ago. What was the purchase price of the office building five years ago?

Strategy

To find the price five years ago:

* Assign a variable to the purchase price of the office building five years ago.

* Write two expressions for the same value.

* Write and solve an equation.

Solution

price 5 years ago: p

$\boxed{\text{the purchase price today}}$ is $\boxed{\text{2.5 times the purchase price 5 years ago}}$

$$250,000 = 2.5p$$
$$\frac{250,000}{2.5} = \frac{2.5p}{2.5}$$
$$100,000 = p$$

The purchase price five years ago was $100,000.

You Try It 2

As a result of depreciation, the value of a car one year after its purchase was $9500. This is 0.80 times its original value. Find the original value of the car.

Your strategy

Your solution

Solution on p. A17

EXERCISE 5.3A

Solve.

1. A mixture contains three times as many peanuts as cashews. Express the amount of peanuts in the mixture in terms of the amount of cashews in the mixture.

2. The dividend paid on a company's stock is one-twentieth of the price of the stock. Express the dividend paid on the stock in terms of the price of the stock.

3. A coffee merchant used mocha java beans and espresso beans in making a coffee blend. There are 35 more pounds of mocha java beans in the blend than espresso beans. Express the amount of mocha java beans in the blend in terms of the amount of espresso beans in the blend.

4. A Burger King restaurant served 145 fewer customers on Thursday of one week than on Friday of the same week. Express the number of customers served on Thursday in terms of the number of customers served on Friday.

5. A K mart store is open 4 fewer hours on Sunday than on Saturday. Express the number of hours the store is open on Sunday in terms of the number of hours it is open on Saturday.

6. A shoe store orders 8 times as many women's size 7 Nikes as women's size 11 Nikes. Express the number of Nikes ordered in women's size 7 in terms of the number ordered in women's size 11.

7. An AT&T customer spent $18.75 more on long-distance phone calls this month than last month. Express the amount the customer spent on long-distance phone calls this month in terms of the amount spent on long-distance phone calls last month.

8. A Chrysler showroom had 16 more requests for a test drive during the weekend than it had during the weekdays. Express the number of requests for a test drive during the weekend in terms of the number of requests for a test drive during the weekdays.

9. The Consumer Price Index for energy in 1990 was 4 times the Consumer Price Index for energy in 1970. Express the Consumer Price Index for energy in 1990 in terms of the Consumer Price Index for energy in 1970.

10. The Macintosh Plus computer screen displays 170 more pixels horizontally than it displays vertically. Express the number of pixels the computer displays horizontally in terms of the number of pixels it displays vertically.

1. _____

2. _____

3. _____

4. _____

5. _____

6. _____

7. _____

8. _____

9. _____

10. _____

EXERCISE 5.3B

Solve.

11. A high-density computer disk can store approximately 1,400,000 bytes of data. This is 5 times as much as a low-density disk can store. How many bytes can be stored on a low-density disk?

12. The height of a computer monitor screen is 15 inches. This is the same as 0.75 times the length of the screen. What is the length of the computer monitor screen?

13. The per capita personal income in the United States in 1990 was $18,696. This is $8786 more than the per capita personal income in 1980. What was the per capita personal income in the United States in 1980?

14. The median family income in the United States in 1990 was $33,956. This is $12,933 more than the median family income in 1980. What was the median family income in the United States in 1980?

15. In 1990, on average the U.S. consumer spent $1910 on durable goods. This equals $2838 less than the average spent by a U.S. consumer on nondurable goods. What was the average amount spent on nondurable goods by a U.S. consumer in 1990?

16. In 1991 the number of new housing starts in the United States was 1,014,000. This is 179,000 fewer housing starts than in 1990. Find the number of housing starts in the United States in 1990.

17. In 1990 the number of passenger cars made and sold in the United States was 5,502,000. This represents approximately 10 times the number of passenger cars made in the United States and sold in foreign countries. How many passenger cars were made in the United States and sold in foreign countries in 1990?

18. The circulation of *TV Guide* is 15,000,000. This amounts to 3 times the circulation of *Family Circle*. What is the circulation of *Family Circle*?

19. In 1991 Coca-Cola's sales totaled $11,571.6 million. This equals $8199.6 million less than Pepsico's total sales. Find Pepsico's total sales in 1991.

20. In 1991 the operating revenues of United Parcel Service were $15,047.4 million. This represents $7359.1 million more than the operating revenues of Federal Express in 1991. What were the operating revenues of Federal Express in 1991?

11. _____

12. _____

13. _____

14. _____

15. _____

16. _____

17. _____

18. _____

19. _____

20. _____

CALCULATORS

Solving Equations

In this chapter, the following rules for solving equations were presented. When a number is **added** to a variable, **subtract** the number from each side of the equation. When a number is **subtracted** from a variable, **add** the number to each side of the equation. When a variable is **multiplied** by a number, **divide** each side of the equation by the number.

Consider the equation $x + 4.5 = 12$.
4.5 is *added* to the variable x. To solve the equation, it is necessary to *subtract* 4.5 from each side of the equation.

$$x + 4.5 = 12$$
$$x + 4.5 - 4.5 = 12 - 4.5$$
$$x = 7.5$$

The solution can be found by performing the operation $12 - 4.5$ on a calculator.

$$12 - 4.5 = 7.5$$

To solve the equation $y - 8.7 = 5$ using a calculator, enter the 5 (the number that is alone on one side of the equation). Because 8.7 is *subtracted* from the variable, *add* it to 5 ($5 + 8.7$). The solution is 13.7.

To solve the equation $7z = 280$ using a calculator, enter the 280 (the number that is alone on one side of the equation). Because 7 *multiplies* the variable, *divide* 280 by 7 ($280 \div 7$). The solution is 40.

What operations should be performed on a calculator in order to solve the following equations?

1. $7.5 + A = 9.4$ 2. $6.1 = T - 8.2$ 1. _____   2. _____

3. $6.3 = 2.1c$ 4. $300 = B(0.5)$ 3. _____   4. _____

5. $0.72 = 0.9 + d$ 6. $M - 32.5 = 79.9$ 5. _____   6. _____

BUSINESS CASE STUDY

Break-Even Analysis

The costs involved in manufacturing a product can be broadly classified as either fixed costs or variable costs. **Fixed costs** are those costs that the company has to pay no matter how many units of the products are produced. Rent is an example of a fixed cost; it remains unchanged despite a change in the production level. **Variable costs** are costs that depend entirely on the number of units produced. A company's cost for the materials used to make chairs is a variable cost; the more chairs produced, the higher the cost of the materials.

Break–even analysis is a method used to determine the sales volume required for a company to break even—that is, to experience neither a profit nor a loss on the sale of its product. The break–even point represents the number of units that must be made and sold in order for income from sales to equal the cost of production. Two formulas associated with the break–even point are:

$$B = \frac{F}{S - V} \quad \text{and} \quad S = \frac{F + BV}{B}$$

where B is the break–even point, F is the total fixed costs, S is the selling price per unit, and V is the total variable costs per unit.

You are a marketing manager for a manufacturing company that produces rocking chairs. Sales have been declining during the past year. Management, concerned about breaking even during the coming year, has asked you to prepare some figures. You have spoken to the company's accountant and learned that the company's fixed costs are $24,000 and that the total variable costs per chair are $80.

1. You want to determine the break–even point when the selling price of a chair is $230. Which formula given above would be easiest to use for this purpose? What is the break–even point? Express the answer in number of chairs. How can you express this break–even point in terms of total sales? What would the total sales be?
2. You want to determine what the selling price of a rocking chair should be in order for the break–even point to be 120 chairs. Which formula given above would be easiest to use for this purpose? What is the selling price of a rocking chair when the break–even point is 120 chairs?
3. During the past year, the company sold 130 rocking chairs at a selling price of $250 per chair. Did the company lose money, break even, or earn a profit?

CHAPTER SUMMARY

Key Words

A **variable** is a letter of the alphabet used to stand for a quantity that is unknown. (Objective 5.1A)

An **equation** states that two expressions are equal. A **solution** of an equation is a number that, when substituted for the variable, results in a true equation. (Objective 5.1A)

To **solve an equation** means to find the solution of the equation. In solving an equation, the goal is to get the variable alone on one side of the equation; the other side of the equation is the solution. (Objective 5.1B)

When translating a word problem into an equation, remember that the following words or phrases translate to "=": *equals, is, represents, amounts to, totals, is the same as.* (Objective 5.3B)

Essential Rules

To determine **whether a given number is a solution of an equation,** replace the variable in the equation with the given number. Then simplify. If the result is a true equation, the given number is a solution of the equation. (Objective 5.1A)

Properties of Equations

The same number can be added to or subtracted from each side of an equation without changing the solution of the equation.

Each side of an equation can be multiplied by or divided by the same number (except 0) without changing the solution of the equation. (Objective 5.1B)

To solve an equation in which a number is added to a variable, subtract the number from each side of the equation. (Objective 5.1B)

To solve an equation in which a number is subtracted from a variable, add the number to each side of the equation. (Objective 5.1B)

To solve an equation in which a variable is multiplied by a number, divide each side of the equation by the number. (Objective 5.1B)

To solve a problem that involves a formula, substitute the known quantities into the given formula. Then solve the resulting equation. (Objective 5.2A)

REVIEW / TEST

1. Is 4.7 a solution of the equation
   $d + 3.92 = 8.62$?

2. Is 2.3 a solution of the equation
   $10.9 = 4.7t$?

3. Solve: $M = 247.50 - 184.75$

4. Solve: $F - 18 = 32$

5. Solve: $0.90R = 18$

6. Solve: $4.9 + C = 8.7$

7. Solve: $11.2 = 6.3 + b$

8. Solve: $48 = 0.75S$

9. Solve: $2.46 = P - 8.19$

10. Solve: $1.36 = N + 0.59$

11. Solve: $\dfrac{3}{8} + n = \dfrac{5}{8}$

12. Solve: $72 = B(0.45)$

1. _____

2. _____

3. _____

4. _____

5. _____

6. _____

7. _____

8. _____

9. _____

10. _____

11. _____

12. _____

13. _____

14. _____

Solve.

13. Henry Poulin must pay a total of $2760 in monthly payments of $345. In how
    many months will the $2760 debt be paid off? Use the formula $A = P \cdot N$, where
    A is the amount to be repaid, P is the monthly payment, and N is the number of
    months.

14. The value of a stock investment you hold is $2800. The original value of the
    investment one year ago was $2150. Find the increase in the value of your
    investment. Use the formula $A = P + I$, where A is the value of the investment
    after one year, P is the original investment, and I is the increase in value of the
    investment.

15. Phil Aaronsen made a down payment of $18,900 on a home. He took out a mortgage loan on the house for $75,600. What was the selling price of the house? Use the formula $M = S - D$, where M is the mortgage loan amount, S is the selling price, and D is the down payment.

16. If it costs you $.81 to operate an 1800–watt air conditioner for 5 hours, what is your cost per kilowatt hour? Use the formula $c = \frac{1}{1000}wtk$, where c is the cost of operating an appliance, w is the number of watts, t is the time in hours, and k is the cost per kilowatt–hour.

17. The circulation of the Sunday edition of the *New York Times* is 570,000 more than the circulation of that newspaper's weekday morning edition. Express the circulation of the Sunday edition of the *New York Times* in terms of the circulation of the weekday morning edition.

18. The amount of freight carried by railroads in the United States is 10 times the amount of freight carried by air carriers. Express the amount of freight carried by railroads in the United States in terms of the amount of freight carried by air carriers.

19. The per capita property tax in the United States in 1990 was $626. This was $324 more than the per capita property tax in the United States in 1980. What was the per capita property tax in the United States in 1980?

20. In 1991 K mart had sales totaling $34,969 million. This was approximately twice J.C. Penney's total sales. Find J.C. Penney's total sales in 1991.

15. _____

16. _____

17. _____

18. _____

19. _____

20. _____

6

Percent and Business Applications

A

N INDEX NUMBER measures the change in a quantity, such as cost, over a period of time. One of the most widely used indexes is the **consumer price index (CPI)**, or the "cost-of-living index." The CPI, which includes the selling prices of key consumer goods and services, indicates the relative change in the price of these items.

The years 1982–1984 are used as the base years for the consumer price index; the CPI for these years is 100. This means that a consumer good that cost $100 in the period from 1982 to 1984 and has a CPI of 123 this year would cost $123 today.

The CPI of different goods and services—such as food, transportation, medical care, and apparel—are available for a given year. There is also one average annual figure. For example, the CPI for all items in 1992 was 140.3; the CPI for shelter in 1992 was 151.2.

An index number is actually a percent written without a percent sign. A CPI for 1992 of 140.3 means that the average price of a consumer good in 1992 was 140.3% of its price in 1982–1984. The average price for shelter in 1992 was 151.2% of its price in 1982–1984. These figures show that the price of shelter was rising more rapidly than the price of most consumer goods and services.

Closely related to the CPI is inflation. **Inflation** is an economic condition during which there are increases in the cost of goods and services. Inflation is expressed as a percent; for example, we speak of an annual inflation rate of 7%.

But does an annual inflation rate of 7% mean that the price of all goods and services rose 7% during the year? No. We noted that for 1992, the increase in the price of shelter was greater than the average increase in prices. The 1992 CPI for fuel oil was 90.7. This means that in 1992, a consumer could purchase for $90.70 the same amount of fuel oil that $100 would have purchased in the period 1982 to 1984.

The CPI for 1991 was 136.2, and the CPI for 1992 was 140.3. Therefore, prices were increasing in 1992. But areas of more rapid increase in price, such as shelter, were offset by areas of price decreases, such as in fuel oil.

For 1991 the CPI for shelter was 146.3, and the CPI for fuel oil was 94.6. Compare these figures with the 1992 figures. Did the price of these items increase or decrease during 1992?

Converting Decimals, Fractions, and Percents

Objective 6.1A   *To write a percent as a fraction or a decimal*

"A manufacturer's discount of 25%" and "an 8% increase in pay" are examples of the many ways in which percent is used in business. **Percent** means "parts of 100." Thus 27% means 27 parts of 100.

In business problems involving a percent, it is usually necessary either to rewrite the percent as a fraction or a decimal, or to rewrite a fraction or a decimal as a percent.

To write a percent as a fraction, drop the percent sign and multiply by $\frac{1}{100}$.

Write 27% as a fraction.

◆ Drop the percent sign and multiply by $\frac{1}{100}$. $27\% = 27\left(\frac{1}{100}\right) = \frac{27}{100}$

To write a percent as a decimal, drop the percent sign and multiply by 0.01.

Write 33% as a decimal.

◆ Drop the percent sign and multiply by 0.01. $33\% \quad = \quad 33(0.01) \quad = \quad 0.33$

> Move the decimal point two places to the left and drop the percent sign.

Example 1

Write 120% as a fraction and as a decimal.

Solution

$120\% = 120\left(\frac{1}{100}\right) = \frac{120}{100} = 1\frac{1}{5}$

$120\% = 120(0.01) = 1.2$

Note that percents larger than 100 are greater than 1.

You Try It 1

Write 125% as a fraction and as a decimal.

Your solution

Example 2

Write $16\frac{2}{3}\%$ as a fraction.

Solution

$16\frac{2}{3}\% = 16\frac{2}{3}\left(\frac{1}{100}\right) = \frac{50}{3}\left(\frac{1}{100}\right) = \frac{1}{6}$

You Try It 2

Write $33\frac{1}{3}\%$ as a fraction.

Your solution

Solutions on p. A18

Objective 6.1B *To write a fraction or a decimal as a percent*

A fraction or a decimal can be written as a percent by multiplying by 100%.

Write $\frac{5}{8}$ as a percent.

♦ Multiply by 100%.

$$\frac{5}{8} = \frac{5}{8}(100\%) = \frac{500}{8}\% = 62.5\%, \text{ or } 62\frac{1}{2}\%$$

Write 0.82 as a percent.

♦ Multiply by 100%.

$$0.82 \quad = \quad 0.82(100\%) \quad = \quad 82\%$$

> Move the decimal point two places to the right and write the percent sign.

Example 3

Write 0.015 as a percent.

Solution

$0.015 = 0.015(100\%) = 1.5\%$

You Try It 3

Write 0.048 as a percent.

Your solution

Example 4

Write $0.33\frac{1}{3}$ as a percent.

Solution

$0.33\frac{1}{3} = 0.33\frac{1}{3}(100\%) = 33\frac{1}{3}\%$

You Try It 4

Write $0.62\frac{1}{2}$ as a percent.

Your solution

Example 5

Write $\frac{2}{3}$ as a percent. Write the remainder in fractional form.

Solution

$\frac{2}{3} = \frac{2}{3}(100\%) = \frac{200}{3}\% = 66\frac{2}{3}\%$

You Try It 5

Write $\frac{5}{9}$ as a percent. Round to the nearest tenth of a percent.

Your solution

Solutions on p. A18

EXERCISE 6.1A

Write as a fraction and as a decimal.

1. 25%   **2.** 40%   **3.** 130%   **4.** 150%   **5.** 100%

6. 425%   **7.** 70%   **8.** 55%   **9.** 45%   **10.** 90%

11. 64%   **12.** 32%   **13.** 8%   **14.** 6%   **15.** 1%

Write as a fraction.

16. $66\frac{2}{3}\%$ **17.** $12\frac{1}{2}\%$ **18.** $3\frac{1}{8}\%$ **19.** $11\frac{1}{9}\%$ **20.** $37\frac{1}{2}\%$

Write as a decimal.

21. 6.5%   **22.** 8.25%   **23.** 12.3%   **24.** 80.4%   **25.** 2%

1. _____
2. _____
3. _____
4. _____
5. _____
6. _____
7. _____
8. _____
9. _____
10. _____
11. _____
12. _____
13. _____
14. _____
15. _____
16. _____
17. _____
18. _____
19. _____
20. _____
21. _____
22. _____
23. _____
24. _____
25. _____

EXERCISE 6.1B

Write as a percent.

26.	0.16	27.	0.73	28.	0.59	29.	0.82	30.	0.05

31.	0.01	32.	0.7	33.	0.9	34.	0.2	35.	0.8

36.	1.24	37.	1.37	38.	1.1	39.	0.004	40.	0.006

Write as a percent. Round to the nearest tenth of a percent.

41. $\frac{37}{100}$ 42. $\frac{1}{3}$ 43. $\frac{2}{5}$ 44. $\frac{5}{8}$ 45. $\frac{1}{6}$

Write as a percent. Write the remainder in fractional form.

46. $\frac{1}{3}$ 47. $\frac{1}{12}$ 48. $\frac{2}{9}$ 49. $1\frac{2}{3}$ 50. $\frac{1}{8}$

26. _____

27. _____

28. _____

29. _____

30. _____

31. _____

32. _____

33. _____

34. _____

35. _____

36. _____

37. _____

38. _____

39. _____

40. _____

41. _____

42. _____

43. _____

44. _____

45. _____

46. _____

47. _____

48. _____

49. _____

50. _____

The Percent Equation

Objective 6.2A   *To solve the basic percent equation*

Solving a problem that involves a percent requires solving the basic percent equation.

The Basic Percent Equation

$$\text{Part} = \text{base} \times \text{rate}$$
$$P \;=\; B \;\times\; R$$

In any percent problem, two elements of the basic percent equation are given and one element is unknown.

When translating a problem involving a percent into an equation, remember that:

The rate is the percent.

The base usually follows the word "of."

Consider the problem "45% of 320 is what number?"

45%	of	320	is	what number?
45% is the **rate.** $45\% = 0.45$ Substitute 0.45 for R.		320 follows the word "of." It is the **base.** Substitute 320 for B.		The **part** is unknown. It will be the variable in the equation.

This problem is solved as shown below.

Write the basic percent equation. $\text{Part} = \text{base} \times \text{rate}$
$$P = B \times R$$

Substitute 0.45 for R and 320 for B. $P = 320 \times 0.45$
Solve for P. $P = 144$

45% of 320 is 144.

20% of what number is 30?

Given: $P = 30$
$R = 20\% = 0.20$

Unknown: B

$$P = B \times R$$
$$30 = B(0.20)$$
$$\frac{30}{0.20} = \frac{B(0.20)}{0.20}$$
$$150 = B$$

20% of 150 is 30.

Find 25% of 200.

Given: $B = 200$

$R = 25\% = 0.25$

Unknown: P

$P = B \times R$

$= 200(0.25)$

$= 50$

25% of 200 is 50.

What percent of 40 is 30?

Given: $P = 30$

$B = 40$

Unknown: R

The solution must be written as a percent to answer the question.

$P = B \times R$

$30 = 40 \cdot R$

$\dfrac{30}{40} = \dfrac{40 \cdot R}{40}$

$0.75 = R$

$75\% = R$

30 is 75% of 40.

In most cases, we write the percent as a decimal before solving the basic percent equation. However, some percents are more easily written as a fraction. For example,

$$33\tfrac{1}{3}\% = \tfrac{1}{3} \qquad 66\tfrac{2}{3}\% = \tfrac{2}{3} \qquad 16\tfrac{2}{3}\% = \tfrac{1}{6} \qquad 83\tfrac{1}{3}\% = \tfrac{5}{6}$$

A table of equivalent fractions, decimals, and percents is given on page 177.

Example 1

What is 4% of 600?

Solution

$P = B \times R$

$= 600(0.04)$

$= 24$

4% of 600 is 24.

Example 2

12% of what is 480?

Solution

$P = B \times R$

$480 = B(0.12)$

$\dfrac{480}{0.12} = \dfrac{B(0.12)}{0.12}$

$4000 = B$

12% of 4000 is 480.

You Try It 1

8% of 2000 is what?

Your solution

You Try It 2

215 is 86% of what?

Your solution

Solutions on p. A19

Example 3

What percent of 32 is 16?

Solution

$$P = B \times R$$

$$16 = 32 \cdot R$$

$$\frac{16}{32} = \frac{32R}{32}$$

$$0.5 = R$$

$$50\% = R$$

16 is 50% of 32.

You Try It 3

24 is what percent of 75?

Your solution

Solution on p. A19

Objective 6.2B   *To solve percent problems*

To solve percent problems, it is necessary to identify the part, the base, and the rate.

The rate is the percent. You can identify the rate by the percent symbol.

The base follows the word "of." It is generally the beginning value, or whole, with which we are comparing another value.

The part is the portion of the base that results from the base being multiplied by the rate. Note that the part may not be a number less than the base; if the rate is greater than 100%, the part will be a number that is larger than the base.

Each year Ann Woodman receives a payment that equals 8% of the value of her investment in a company. This year that payment amounted to $640. Find the value of her investment this year.

◆ To find the value of her investment this year, solve the basic percent equation for the base. The rate is 8% = 0.08. The part is 640.

$$P = B \times R$$
$$640 = B(0.08)$$

$$\frac{640}{0.08} = \frac{B(0.08)}{0.08}$$

$$8000 = B$$

This year her investment is worth $8000.

Example 4

A plumbing company sold 6 of the 30 sink basins it had in stock. What percent of the sinks in stock did the company sell?

Strategy

To find the percent sold, solve the basic percent equation for rate. (part = 6, base = 30)

Solution

$$P = B \times R$$

$$6 = 30 \cdot R$$

$$\frac{6}{30} = \frac{30R}{30}$$

$$0.2 = R$$

$$20\% = R$$

20% of the sinks in stock were sold.

Example 5

A business bought a used copy machine for $1650, which was 75% of the original cost. What was the original cost of the copier?

Strategy

To find the original cost, solve the basic percent equation for the base. (part = 1650, rate = 0.75)

Solution

$$P = B \times R$$

$$1650 = B(0.75)$$

$$\frac{1650}{0.75} = \frac{B(0.75)}{0.75}$$

$$2200 = B$$

The original cost of the copier was $2200.

You Try It 4

A company had sales of $450,000 in 1988. In 1993 the sales were $1,350,000. What percent of the 1988 sales were the 1993 sales?

Your strategy

Your solution

You Try It 5

A manufacturer of transistors expects 1.1% of its transistors to be defective. In a batch of 350,000 transistors, how many does the manufacturer expect to be defective?

Your strategy

Your solution

Solutions on p. A19

_E_XERCISE 6.2A

Solve.

1. 12 is what percent of 50?

2. What percent of 125 is 50?

3. Find 18% of 40.

4. What is 25% of 60?

5. 12% of what is 48?

6. 45% of what is 9?

7. What is $33\frac{1}{3}$% of 27?

8. Find $16\frac{2}{3}$% of 30.

9. What percent of 12 is 3?

10. 10 is what percent of 15?

11. 60% of what is 3?

12. 75% of what is 6?

13. 37 is what percent of 148?

14. What percent of 150 is 33?

15. 82 is 20.5% of what?

16. 2.4% of what is 21?

17. What is 6.5% of 200?

18. 12 is what percent of 6?

19. What is 250% of 18?

20. 33 is 220% of what?

21. 160% of what is 40?

22. What percent of 344 is 43?

23. Find 15.4% of 50.

24. What is 18.5% of 46?

25. 3 is 1.5% of what?

26. 0.75% of what is 3?

1. _____

2. _____

3. _____

4. _____

5. _____

6. _____

7. _____

8. _____

9. _____

10. _____

11. _____

12. _____

13. _____

14. _____

15. _____

16. _____

17. _____

18. _____

19. _____

20. _____

21. _____

22. _____

23. _____

24. _____

25. _____

26. _____

EXERCISE 6.2B

Solve.

27. The harvest from a small orange grove was 560 crates. Five percent of the 560 crates were damaged during shipping. How many crates were damaged during shipping?

28. Tomas Gonzalez, who was earning an hourly wage of $12.50, received a 6% raise. Find the increase in Tomas's hourly wage.

29. The R. B. Hayes Company spends 15% of its $35,000 advertising budget for newspaper ads. How much does the company spend for newspaper advertising?

30. During the packaging process for vegetables, spoiled vegetables are discarded by an inspector. In one day an inspector had to discard 50 pounds of the 1250 pounds of vegetables inspected. What percent of the vegetables inspected were spoiled?

31. In testing the strength of nylon rope, an inspector found that 15 pieces of the 250 pieces tested did not meet the standards. What percent of the pieces of nylon rope tested did not meet the standards?

32. A product preference survey revealed that 7 out of the 200 people surveyed liked a new product. What percent of the people tested liked the product?

33. The market price of a share of stock one year ago was $50. Since then, the market price has increased 8%. Find the increase in the market price.

34. During a pre-employment test, a typist made errors on three words on the typing test. This was 2% of the total number of words typed. How many words were typed?

35. Browning Department Store advertised a style of all-leather shoes for $109.25. This was 115% of the price at a competitor's store. Find the price at the competitor's store.

36. Williams Department Store employs 125 people and must hire an additional 20% for the holiday season. What is the total number of employees needed for the holiday season?

37. Bradshaw Enterprises pays its sales managers $4500 per month this year. With a 5% raise, what will a sales manager's monthly salary be next year?

27. _____

28. _____

29. _____

30. _____

31. _____

32. _____

33. _____

34. _____

35. _____

36. _____

37. _____

SECTION 6.3 | Percent Increase and Percent Decrease

Objective 6.3A   *To find percent increase*

Percent increase is used to show how much a quantity has increased over its original value. The statements "car prices will show a 3.5% increase over last year's prices" and "employees were given an 8% pay increase" are illustrations of the use of percent increase.

A company's production increased from 50,000 to 51,500 in one year.

To find the percent increase:

♦ First find the amount of increase.

New value	−	original value	=	amount of increase
51,500	−	50,000	=	1500

♦ Then solve the basic percent equation for rate.

Part = base × rate

Amount of increase	=	original value	×	percent increase

$$1500 = 50{,}000 \cdot R$$
$$\frac{1500}{50{,}000} = \frac{50{,}000R}{50{,}000}$$
$$0.03 = R$$
$$3\% = R$$

The company's production increased by 3% over the previous year.

This year a company's production increased 12% over last year's production of 30,000 units. Find this year's production.

♦ First solve the basic percent equation for part.

Amount of increase	=	original value	×	percent increase

Part	=	base	×	rate
P	=	30,000(0.12)		
	=	3600		

The amount of increase in production was 3600 units.

♦ Then add the amount of increase to the original value.

Original value	+	amount of increase	=	new value
30,000	+	3600	=	33,600

The company's production this year was 33,600 units.

Example 1

The average price of gasoline rose from $.92 to $1.15 in four months. What was the percent increase in the price of gasoline?

Strategy

To find the percent increase:

* Find the amount of increase.

* Solve the basic percent equation for rate. The part is the amount of increase. The base is the original price, $.92.

Solution

$$\begin{matrix} \text{New} \\ \text{value} \end{matrix} - \begin{matrix} \text{original} \\ \text{value} \end{matrix} = \begin{matrix} \text{amount of} \\ \text{increase} \end{matrix}$$

$$\$1.15 - \$.92 = \$.23$$

Part = base × rate

$$0.23 = 0.92 \cdot R$$

$$\frac{0.23}{0.92} = \frac{0.92R}{0.92}$$

$$0.25 = R$$

$$25\% = R$$

The percent increase was 25%.

You Try It 1

An industrial plant increased its number of employees from 2000 to 2250. What was the percent increase in the number of employees?

Your strategy

Your solution

Example 2

Paula Noble was earning a wage of $5.80 per hour before she received a 10% increase in pay. What is Paula's new hourly wage?

Strategy

To find the new hourly wage:

* Solve the basic percent equation for part.

* Add the amount of increase to the original wage.

Solution

Part = base × rate

$$P = 5.80(0.10)$$

$$= 0.58$$

$$\$5.80 + \$.58 = \$6.38$$

The new hourly wage is $6.38.

You Try It 2

Deon Brown was earning a wage of $6.50 per hour before he got a 14% increase in pay. What is Deon's new hourly wage?

Your strategy

Your solution

Solutions on p. A20

*O*bjective 6.3B   *To find percent decrease*

Percent decrease is frequently used to show how much a quantity has decreased from its original value. The statements "the unemployment rate decreased by 0.4% over last month" and "there has been a 12% decrease in the number of industrial accidents" are illustrations of the use of percent decrease.

The price of a business calculator decreased from $60 to $52.80 in one year.

To find the percent decrease:

♦ First find the amount of decrease.

$$\begin{array}{ccccc} \text{Original} & - & \text{new} & = & \text{amount of} \\ \text{value} & & \text{value} & & \text{decrease} \end{array}$$

$$\$60 \quad - \quad \$52.80 \quad = \quad \$7.20$$

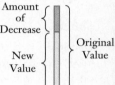

♦ Then solve the basic percent equation for rate.

$$\text{Part} \quad = \quad \text{base} \quad \times \quad \text{rate}$$

$$\begin{array}{ccccc} \text{Amount of} & = & \text{original} & \times & \text{percent} \\ \text{decrease} & & \text{value} & & \text{decrease} \end{array}$$

$$7.20 = 60 \cdot R$$

$$\frac{7.20}{60} = \frac{60R}{60}$$

$$0.12 = R$$

$$12\% = R$$

The price decreased by 12% below the previous year's price.

Today the price of a hand-held calculator is 8% less than last year's price of $39.50. Find the price of the calculator today.

♦ First solve the basic percent equation for part.

$$\begin{array}{ccccc} \text{Amount of} & = & \text{original} & \times & \text{percent} \\ \text{decrease} & & \text{value} & & \text{decrease} \end{array}$$

$$\text{Part} = \text{base} \times \text{rate}$$

$$P = 39.50(0.08)$$
$$= 3.16$$

The amount of decrease in price is $3.16.

♦ Then subtract the amount of decrease from the original value.

$$\begin{array}{ccccc} \text{Original} & - & \text{amount of} & = & \text{new} \\ \text{value} & & \text{decrease} & & \text{value} \end{array}$$

$$39.50 \quad - \quad 3.16 \quad = \quad 36.34$$

Today's price for the calculator is $36.34.

Example 3

Because of unusually high temperatures, the number of people vacationing at a desert resort dropped from 650 in June to 525 in July. Find the percent decrease in vacationers. Round to the nearest tenth of a percent.

Strategy

To find the percent decrease:

* Find the amount of decrease.
* Solve the basic percent equation for rate.

Solution

$$\begin{array}{ccccc} \text{Original} & & \text{new} & & \text{amount of} \\ \text{value} & - & \text{value} & = & \text{decrease} \end{array}$$

$$650 \quad - \quad 525 \quad = \quad 125$$

Part = base × rate

$$125 = 650 \cdot R$$

$$\frac{125}{650} = \frac{650R}{650}$$

$$0.192 \approx R$$

$$19.2\% \approx R$$

The percent decrease was 19.2%.

Example 4

The total sales for December for a stationery store were $26,000. For January, total sales showed an 8% decrease from December's sales. What were the total sales for January?

Strategy

To find the total sales for January:

* Find the amount of decrease by solving the basic percent equation for part.
* Subtract the amount of decrease from the December sales.

Solution

Part = base × rate

$$P = 26,000(0.08)$$

$$= 2080$$

$$26,000 - 2080 = 23,920$$

Total sales for January were $23,920.

You Try It 3

Sales of automobiles at a car dealership dropped from 150 in June to 120 in July. Find the percent decrease in car sales.

Your strategy

Your solution

You Try It 4

The market value of a new truck is reduced by 30% after one year of ownership. Using this estimate, what is the market value of a $28,800 new truck after one year?

Your strategy

Your solution

Solutions on p. A20

EXERCISE 6.3A

Solve.

1. The market value of a $2500 investment increased $500. What percent increase does this represent?

2. A stock that sold for $30 per share increased in market value by $1.50 in one day. What percent increase does this represent?

3. Bay Shore Company's production level increased from 250 units per week to 275 units per week. What percent increase does this represent?

4. The number of management trainees working for a company has increased from 36 to 42. What percent increase does this represent?

5. A manufacturer of radios increased its monthly output of 2000 radios by 15%. What was the amount of increase?

6. A contractor built an addition on a house, increasing the 1250-square-foot home by 20%. How much larger, in square feet, is the home now?

7. A new labor contract calls for an 8.5% increase in pay for all employees. (a) What is the amount of increase for an employee who earns $256 per week? (b) What is the weekly wage of this employee after the wage increase?

8. Lee Won's salary this year is $32,000. His salary will increase by 11% next year. What will Lee's salary be next year?

9. The Fremont Company increased its number of employees from 500 to 525. What percent increase does this represent?

10. A company plans to increase its 5000-per-month production schedule by 6.5%. How many units will be produced each month?

1. _____

2. _____

3. _____

4. _____

5. _____

6. _____

7. (a) _____

   (b) _____

8. _____

9. _____

10. _____

EXERCISE 6.3B

Solve.

11. Last month a company's sales totaled $35,000. This month's sales totaled $28,000. What percent decrease does this represent?

12. Last year a company's travel expenditures totaled $2500. This year the travel expenditures totaled $2300. What percent decrease does this represent?

13. An auto parts store sold 1250 parts in March. In April the store sold 300 fewer parts than in March. What was the percent decrease in the number of parts sold?

14. Last winter, Hilltop Restaurant's utility bill averaged $400 per month. By installing energy-saving equipment, the restaurant reduced its monthly utility bill by $80. What percent decrease does this represent?

15. A ski resort employs 1200 people during the skiing season. At the end of the skiing season, the resort reduces the number of employees by 45%. What is the decrease in the number of employees?

16. A new production method reduced the time needed to clean a piece of metal from 8 minutes to 5 minutes. What percent decrease does this represent?

17. Because of a decrease in demand for black-and-white television sets, a television dealer reduced the orders for these models from 20 per month to 8 per month. What percent decrease does this represent?

18. Last year a company earned a profit of $250,000. This year, because of unexpected losses, the company's profits were 8% less than last year's. What is the profit this year?

19. A sales manager's monthly expense for gasoline was $76. By joining a car pool, the manager was able to reduce this expense by 20%. What is the monthly gasoline bill now?

20. As a result of an increased number of service lines at a grocery store, the average amount of time a customer waits in line has decreased from 3.5 minutes to 2.8 minutes. What percent decrease does this represent?

11. _____

12. _____

13. _____

14. _____

15. _____

16. _____

17. _____

18. _____

19. _____

20. _____

Allocation and Proportion

Objective 6.4A   *To solve allocation problems*

Quantities such as 4 meters, 15 seconds, and 8 dollars are number quantities written with units. In these examples, the units are meters, seconds, and dollars.

A **ratio** is the quotient of two quantities that have the same unit. This comparison can be written as a fraction or as two numbers separated by a colon.

During the month of July, a small business had sales of $15,000 and costs totaling $20,000. The ratio of sales to costs is written

$$\frac{\$15,000}{\$20,000} = \frac{15,000}{20,000} = \frac{3}{4}, \text{ or } 3:4$$ A ratio is in simplest form when the two numbers do not have a common factor. Note that the units are not written.

Ratios are frequently used in business to describe the allocation, or assignment, of profits or losses. For example, profits earned by a partnership may be distributed on the basis of a ratio.

Three attorneys at law share the profits of their firm in the ratio $5:4:3$. Last year the profits were $120,000. Find the shares received by the three partners.

- Add the numbers in the ratio. $5 + 4 + 3 = 12$

- Form fractions by placing each number in the ratio over the sum of the numbers. $\dfrac{5}{12}, \dfrac{4}{12}, \dfrac{3}{12}$

- Multiply each fraction by last year's profit ($120,000). $\dfrac{5}{12}(120,000) = 50,000$

    $\dfrac{4}{12}(120,000) = 40,000$

    $\dfrac{3}{12}(120,000) = 30,000$

The shares received by the three partners were $50,000, $40,000, and $30,000.

When a company comprises two or more departments, some of the company's operating expenses (such as utilities expense or advertising expense) may not be restricted to one department and must be divided among them. This allocation of expenses among the departments of a company is frequently described in terms of percents.

The Madison Company spent $9600 for newspaper advertising last year. Department A's share of the cost is 60%. Department B's share is 40%. Find the amount allocated to each department.

Department A is allocated 60% of the cost: $0.60(9600) = 5760$
Department B is allocated 40% of the cost: $0.40(9600) = 3840$

Department A is allocated $5760. Department B is allocated $3840.

Example 1

Last year the profits for a partnership were $120,000. The two partners share the profits in the ratio 5 : 3. Find the amounts received by the partners.

Strategy

To find the amounts received:

* Add the numbers in the ratio.

* Form fractions by placing each number in the ratio over the sum of the numbers.

* Multiply each fraction by the profits.

Solution

$5 + 3 = 8$ $\dfrac{5}{8}, \dfrac{3}{8}$

$\dfrac{5}{8}(120,000) = 75,000$

$\dfrac{3}{8}(120,000) = 45,000$

The partners received $75,000 and $45,000.

You Try It 1

The two partners in a partnership share the profits of their business in the ratio 3 : 2. Last year the profits were $115,000. Find the amounts received by the two partners.

Your strategy

Your solution

Example 2

The Wilson Company spent $1800 last year to heat its warehouse. Find the amounts allocated to each of the divisions if Division A's share of the cost is 50%, Division B's share is 30%, and Division C's share is 20%.

Strategy

To find the amounts allocated, multiply each division's percent of the cost by the cost.

Solution

Division A: 0.50(1800) = 900

Division B: 0.30(1800) = 540

Division C: 0.20(1800) = 360

Division A is allocated $900, Division B is allocated $540, and Division C is allocated $360.

You Try It 2

The Salem Corporation paid $6700 for utilities for its factory last year. Find the amounts allocated to the company's two production divisions if Division A's share of the cost is 60% and Division B's share is 40%.

Your strategy

Your solution

Solutions on p. A21

Objective 6.4B   *To solve a proportion*

An equation that expresses the equality of two ratios is a **proportion.** An example of a proportion is shown at the right.

$$\frac{4}{6} = \frac{8}{12}$$

In a proportion, the "cross products" are equal. For the proportion $\frac{2}{3} = \frac{8}{12}$,

$$\frac{2}{3} \bowtie \frac{8}{12} \implies 3 \times 8 = 24$$
$$2 \times 12 = 24$$

Sometimes one of the numbers in a proportion is unknown. In this case it is necessary to *solve* the proportion.

Solve the proportion $\frac{4}{n} = \frac{2}{3}$.

Find the cross products.
Solve for *n*.

$$4 \cdot 3 = n \cdot 2$$
$$12 = 2n$$
$$6 = n$$

The solution is 6.

Example 3

Solve the proportion $\frac{5}{9} = \frac{15}{x}$.

Solution

$5 \cdot x = 9 \cdot 15$

$5x = 135$

$x = 27$

The solution is 27.

You Try It 3

Solve the proportion $\frac{12}{n} = \frac{3}{8}$.

Your solution

Example 4

Solve the proportion $\frac{8}{x} = \frac{4}{3}$.

Solution

$8 \cdot 3 = x \cdot 4$

$24 = 4x$

$6 = x$

The solution is 6.

You Try It 4

Solve the proportion $\frac{2}{5} = \frac{n}{10}$.

Your solution

Solutions on p. A21

Objective 6.4C   *To solve application problems using proportions*

Proportions can be used to solve a wide variety of business application problems, including percent problems. The proportion method of solving percent problems is based on writing two ratios. One ratio is the rate ratio, which is written $\frac{rate}{100}$.

The second ratio is the part-to-base ratio, which is written $\frac{part}{base}$. These two ratios form the proportion

$$\frac{\text{rate}}{100} = \frac{\text{part}}{\text{base}}$$

To use the proportion method, first identify the rate, the part, and the base (the base usually follows the phrase "percent of"). In the proportion, the rate is written as a percent without the percent sign.

What is 23% of 45?

Given: $R = 23\%$

$\qquad$ $B = 45$

Unknown: P

$$\frac{23}{100} = \frac{P}{45}$$
$$23(45) = 100 \cdot P$$
$$1035 = 100P$$
$$10.35 = P$$

23% of 45 is 10.35.

12 is 60% of what number?

Given: $P = 12$

$\qquad$ $R = 60\%$

Unknown: B

$$\frac{60}{100} = \frac{12}{B}$$
$$60 \cdot B = 100(12)$$
$$60B = 1200$$
$$B = 20$$

12 is 60% of 20.

A quality control inspector found that 5 of the 250 watches inspected were defective. What percent of the watches inspected were defective?

◆ Write a proportion.

$\qquad$ The part is 5. The base is 250

◆ Solve for R, the rate.

$$\frac{R}{100} = \frac{P}{B}$$
$$\frac{R}{100} = \frac{5}{250}$$
$$R(250) = 100(5)$$
$$250R = 500$$
$$R = 2$$

The rate is 2%.

2% of the watches inspected were defective.

Example 5

Of the $24,500 it collected, a charity organization spent 12% for administrative costs. How much of the money collected was spent for administrative costs?

Strategy

To find the amount spent for administrative costs, write and solve a proportion.
(rate = 12%, base = 24,500)

Solution

$$\frac{12}{100} = \frac{P}{24,500}$$

$$12(24,500) = 100 \cdot P$$

$$294,000 = 100P$$

$$2940 = P$$

The amount spent for administrative costs was $2940.

You Try It 5

A computer system can be purchased for $44,000, which is 55% of the price five years ago. What was the price of the computer system five years ago?

Your strategy

Your solution

Example 6

In a test of 20 antifreeze solutions, 3 solutions froze above an acceptable temperature. What percent of the solutions did not freeze above the acceptable temperature?

Strategy

To find the percent that did not freeze above the acceptable temperature:

◆ Find the number that did not freeze above the acceptable temperature (20 − 3).

◆ Write and solve a proportion.

Solution

20 − 3 = 17, the number that did not freeze above the acceptable temperature

$$\frac{R}{100} = \frac{17}{20}$$

$$R \cdot 20 = 100(17)$$

$$20R = 1700$$

$$R = 85$$

85% of the solutions did not freeze above the acceptable temperature.

You Try It 6

In a test of 1200 ball-point pens manufactured by a company, 24 were found to be defective. What percent of the pens were not defective?

Your strategy

Your solution

Solutions on p. A21

Example 7

A bank requires a loan payment of $28.35 each month for every $1000 borrowed. At this rate, find the monthly payment for a $6000 loan.

Strategy

To find the monthly payment, write and solve a proportion using P to represent the monthly payment.

Solution

$$\frac{28.35}{1000} = \frac{P}{6000}$$

$$28.35(6000) = 1000 \cdot P$$

$$170,100 = 1000P$$

$$170.10 = P$$

The monthly payment is $170.10.

You Try It 7

The two partners in a partnership share the profits of their business in the ratio 3 : 2. If the partner receiving the larger amount of this year's profits receives $24,000, what amount does the other partner receive?

Your strategy

Your solution

Example 8

An investment of $5000 earns $600 each year. At the same rate, how much money must be invested to earn $900 each year?

Strategy

To find the amount of money that must be invested, write and solve a proportion using x to represent the amount of money.

Solution

$$\frac{600}{5000} = \frac{900}{x}$$

$$600 \cdot x = 5000 \cdot 900$$

$$600x = 4,500,000$$

$$x = 7500$$

To earn $900 each year, $7500 must be invested.

You Try It 8

A chef estimates that 30 pounds of meat will serve 100 customers. Using this estimate, determine how many pounds of meat will be needed to serve 170 customers.

Your strategy

Your solution

Solutions on p. A22

*E*XERCISE 6.4A

Solve.

1. Last year the profits for a partnership were $140,000. The two partners share the profits in the ratio 4:3. Find the amount received by each.

2. The two partners in a partnership share the profits of their business in the ratio 5:3. Last year the profits were $180,000. Find the amounts received by the two partners.

3. Three attorneys share the profits of their firm in the ratio 3:2:1. Last year the profits were $180,000. Find the amount received by each.

4. The profits of a firm are shared by its three partners in the ratio 7:5:3. Last year's profits were $150,000. Find the amount received by each.

5. The Randolph Company spent $35,000 for advertising last year. Department A's share of the cost is 30%. Department B's share is 70%. Find the amount allocated to each department.

6. The Baxter Corporation spent $25,000 for radio advertising last year. Department A's share of the cost is 25%, and Department B's share is 75%. Find the amount allocated to each department.

7. Rent paid on the Albion Company's office space is allocated to the company's three divisions as follows: 20%, 35%, 45%. The company pays $1200 for rent of the office space each month. Find the amount allocated to each of the three divisions.

8. The insurance expense of the Simmons Company is allocated to the company's three departments as follows: 25%, 35%, 40%. The company paid $220,000 for insurance this year. Find the amount allocated to each of the three departments.

9. Of the $8500 spent for newspaper advertising by the Norman Corporation, $3400 is allocated to Department A and $5100 to Department B. What percent of the amount spent for newspaper advertising is allocated to each of the two departments?

10. Of the $4840 spent for utilities by the Broadbent Company, $1694 is allocated to Department A and $3146 to Department B. What percent of the cost is allocated to each of the two departments?

1. _____

   _____

2. _____

   _____

3. _____

   _____

   _____

4. _____

   _____

   _____

5. _____

   _____

6. _____

   _____

7. _____

   _____

   _____

8. _____

   _____

   _____

9. _____

   _____

10. _____

    _____

11. In an accounting firm, one accountant invested $20,000 and a second accountant invested $15,000. Profits are shared in the same ratio as each accountant's investment to the total amount invested. What amount should each accountant receive in a year when profits are $84,000?

12. One fashion designer invested $35,000 in a partnership, and a second designer invested $25,000 in the partnership. Profits are shared in the same ratio as each partner's investment to the total amount invested. What amount should each designer receive in a year when the profits are $78,000?

13. The Jackson Company rents 50,000 square feet of floor space and allocates the yearly rent expense of $32,000 on the basis of floor space occupied by each department. Department A occupies 20,000 square feet, and Department B occupies 30,000 square feet. Find the amount of rent expense allocated to each department.

14. The Washington Company rents 35,000 square feet of floor space and allocates the yearly rent expense of $24,000 on the basis of the floor space occupied by each department. Department A occupies 14,000 square feet, and Department B occupies 21,000 square feet. Find the amount of rent expense allocated to each department.

EXERCISE 6.4B

Solve.

15. $\dfrac{x}{4} = \dfrac{6}{8}$

16. $\dfrac{x}{7} = \dfrac{9}{21}$

17. $\dfrac{12}{18} = \dfrac{x}{9}$

18. $\dfrac{7}{21} = \dfrac{35}{x}$

19. $\dfrac{6}{n} = \dfrac{24}{36}$

20. $\dfrac{3}{n} = \dfrac{15}{10}$

21. $\dfrac{9}{4} = \dfrac{18}{n}$

22. $\dfrac{7}{15} = \dfrac{21}{n}$

23. $\dfrac{x}{12} = \dfrac{5}{6}$

24. $\dfrac{3}{5} = \dfrac{x}{10}$

25. $\dfrac{18}{x} = \dfrac{9}{5}$

26. $\dfrac{n}{11} = \dfrac{32}{4}$

27. $\dfrac{3}{n} = \dfrac{5}{2}$

28. $\dfrac{2}{5} = \dfrac{n}{4}$

29. $\dfrac{n}{6} = \dfrac{2}{15}$

30. $\dfrac{10}{3} = \dfrac{5}{x}$

11. _____

_____

12. _____

_____

13. _____

_____

14. _____

_____

15. _____

16. _____

17. _____

18. _____

19. _____

20. _____

21. _____

22. _____

23. _____

24. _____

25. _____

26. _____

27. _____

28. _____

29. _____

30. _____

EXERCISE 6.4C

Solve.

31. A car dealer offers new-car buyers a 7% rebate on some new-car models. What rebate would a new-car buyer receive on a $22,500 car?

32. A test of the breaking strength of concrete slabs for house foundations found that 3 of the 250 slabs tested did not meet safety requirements. What percent of the slabs tested did not meet safety requirements?

33. The Shelby Company purchased a used copy machine for $825, which was 55% of the original price. What was the original price of the copier?

34. A company had sales of $450,000 in 1983. In 1993 the sales were $1,350,000. What percent of the 1983 sales are the 1993 sales?

35. The price of a videotape recorder decreased from $450 to $360 in one year. Find the percent decrease.

36. A life insurance policy costs $6.87 for every $1000 of insurance coverage. At this rate, what is the cost for $15,000 of insurance coverage?

37. A bank requires a loan payment of $10.84 each month for every $1000 borrowed. At this rate, what is the monthly payment for a $5000 loan?

38. A quality control inspector found 2 defective electric blenders in a shipment of 100 blenders. At this rate, how many blenders would be defective in a shipment of 500?

39. An automobile recall was based on engineering tests that showed 38 steering defects in 1000 cars. At this rate, how many defects would be found in 25,000 cars?

40. A management consulting firm recommends that the ratio of middle-management salaries to management trainee salaries be 5:4. Using this recommendation, what is the yearly middle-management salary when the management trainee yearly salary is $18,000?

31. _____

32. _____

33. _____

34. _____

35. _____

36. _____

37. _____

38. _____

39. _____

40. _____

ata processing manager estimates that the ratio of research computer time to administrative computer time is 2:3. During a month in which the computer was used 200 hours for research, how many hours of computer time were used for administration?

42. The profits of a firm are shared by its two partners in the ratio 7:5. If the partner receiving the larger amount of this year's profits receives $28,000, what amount does the other partner receive?

43. The two partners in a partnership share the profits of their business in the ratio 4:3. If the partner receiving the smaller amount of this year's profits receives $24,000, what amount does the other partner receive?

44. An investment of $6000 earns $500 each year. At this rate, how much money must be invested to earn $800 each year?

45. A chef estimates that 50 pounds of vegetables will serve 130 people. Using this estimate, determine how many pounds of vegetables will be needed to serve 156 people.

46. A caterer estimates that 12 L of fruit punch will serve 40 people. How much punch is necessary to serve 60 people?

47. A farmer estimates that 5600 bushels of wheat can be harvested from 160 acres of land. Using this estimate, determine how many acres are needed to harvest 8120 bushels of wheat.

48. A quality control inspector found that 1.2% of 2500 telephones inspected were defective. How many phones were not defective?

49. A company survey revealed that 57% of its 1500 employees favored a new retirement program. How many employees did not favor the new program?

50. Defects were found in 545 diodes. This was 0.1% of the total number of diodes produced in one week. Find the total number of diodes produced in that week.

41. _____

42. _____

43. _____

44. _____

45. _____

46. _____

47. _____

48. _____

49. _____

50. _____

Table of Equivalent Fractions, Decimals, and Percents

Fraction	Decimal	Percent	Fraction	Decimal	Percent
$\frac{1}{2}$	0.5	50%	$\frac{1}{9}$	$0.11\frac{1}{9}$	$11\frac{1}{9}\%$
$\frac{1}{3}$	$0.33\frac{1}{3}$	$33\frac{1}{3}\%$	$\frac{1}{10}$	0.1	10%
$\frac{2}{3}$	$0.66\frac{2}{3}$	$66\frac{2}{3}\%$	$\frac{3}{10}$	0.3	30%
$\frac{1}{4}$	0.25	25%	$\frac{7}{10}$	0.7	70%
$\frac{3}{4}$	0.75	75%	$\frac{9}{10}$	0.9	90%
$\frac{1}{5}$	0.2	20%	$\frac{1}{12}$	$0.08\frac{1}{3}$	$8\frac{1}{3}\%$
$\frac{2}{5}$	0.4	40%	$\frac{5}{12}$	$0.41\frac{2}{3}$	$41\frac{2}{3}\%$
$\frac{3}{5}$	0.6	60%	$\frac{7}{12}$	$0.58\frac{1}{3}$	$58\frac{1}{3}\%$
$\frac{4}{5}$	0.8	80%	$\frac{1}{15}$	$0.06\frac{2}{3}$	$6\frac{2}{3}\%$
$\frac{1}{6}$	$0.16\frac{2}{3}$	$16\frac{2}{3}\%$	$\frac{1}{16}$	0.0625	6.25%
$\frac{5}{6}$	$0.83\frac{1}{3}$	$83\frac{1}{3}\%$	$\frac{1}{20}$	0.05	5%
$\frac{1}{8}$	0.125	12.5%	$\frac{1}{25}$	0.04	4%
$\frac{3}{8}$	0.375	37.5%	$\frac{1}{50}$	0.02	2%
$\frac{5}{8}$	0.625	62.5%			
$\frac{7}{8}$	0.875	87.5%			

CALCULATORS

Δ% **Key and Decimal Place Selector**

The symbols "Δ%" mean change in percent. The Δ% key on a calculator is used to calculate percent increase and percent decrease. A percent decrease is displayed with a negative sign (–).

A business calculator enables you to determine the number of decimal places to be displayed in the answer to a calculation. The directions for completing the following table read "round to the nearest tenth." Set the format on your calculator to one decimal place. The answer in the display is the rounded value.

Use a calculator to complete the following table. Round the percent to the nearest tenth of a percent.

Product	Cost Last Year	Cost This Year	Amount of Increase or Decrease	Percent Change
File Cabinet	95	99	1.	2.
Secretary's Chair	84	89	3.	4.
Personal Computer	2499	2199	5.	6.
Hard Disk	699	599	7.	8.
Computer Workstation	529	579	9.	10.
Laser Printer	2799	2599	11.	12.
Leather Chair	459	495	13.	14.
Wing Chair	189	199	15.	16.
Conference Table	149	159	17.	18.
Electronic Typewriter	268	259	19.	20.
Mobile Phone	999	799	21.	22.
Cash Register	348	359	23.	24.
Fax Machine	1499	1299	25.	26.

1. _____
2. _____
3. _____
4. _____
5. _____
6. _____
7. _____
8. _____
9. _____
10. _____
11. _____
12. _____
13. _____
14. _____
15. _____
16. _____
17. _____
18. _____
19. _____
20. _____
21. _____
22. _____
23. _____
24. _____
25. _____
26. _____

Business Case Study

Performance Reviews

You are a district manager of the sales division of a medical supplies company. You are the supervisor of five sales representatives who call on hospitals and doctors' offices to sell the products produced by the company.

Each of the five sales representatives in your district is paid a base annual salary. The representatives also earn commissions based on the company commission schedule.

You must prepare the performance reviews for the five representatives you supervise. It is your responsibility to evaluate the performance of each individual and determine each representative's base annual salary for next year. The president of the company has decreed that the total increase in salaries cannot exceed 6% of the total of the present salaries. You could give an across-the-board increase of 6%, meaning that each sales representative's increase would be 6% of her or his base annual salary for this year, but you would prefer to reward those with outstanding performance, giving them a larger raise than the average.

You have compiled the following data on the five sales representatives in your district.

Sales Representative	Number of Years with the Company	Current Base Annual Salary	Total Sales Last Year	Total Sales This Year	Total of Expenses Submitted
A. McNeil	2	$12,475	$276,980	$685,668	$7,678
G. Santiago	3	$13,225	$447,923	$473,306	$3,228
M. Kawn	9	$20,490	$728,274	$545,824	$15,498
B. Mennea	15	$39,330	$439,878	$441,198	$9,244
S. Masters	18	$47,595	$1,494,063	$1,023,889	$22,655

1. What is the total of the current base annual salaries of the five sales representatives?

2. What is the maximum amount of money you can give in salary increases for next year?

3. What is the largest possible total of the annual salaries for the five sales representatives next year?

4. What increases in salary would you give to each of the sales representatives? Prepare a schedule of both the dollar amount of increase and the percent increase for each representative. Defend your decisions. *Remember:* the total of the increases must remain within 6% of the total of the present base annual salaries.

5. Is there any additional information you might like to have on each of the sales representatives before making a decision concerning their increases in salary? If so, what information would you request?

6. Assume the president's mandate concerning salary increases is a reflection of the company "tightening its belt" in all areas of operation. How might this affect your performance reviews for each of the sales representatives?

CHAPTER SUMMARY

Key Words   **Percent** means "parts of 100." Thus 43% means 43 parts of 100. (Objective 6.1A)

Percent increase indicates how much a quantity has increased over its original value. (Objective 6.3A)

Percent decrease indicates how much a quantity has decreased from its original value. (Objective 6.3B)

A **ratio** is the quotient of two quantities that have the same unit. The comparison can be written as a fraction or as two numbers separated by a colon. (Objective 6.4A)

Allocation is the division of expenses among departments or the assignment of profits and losses. (Objective 6.4A)

An equation that expresses the equality of two ratios is a **proportion.** (Objective 6.4B)

Essential Rules   **To write a percent as a fraction,** drop the percent sign and multiply by $\frac{1}{100}$. (Objective 6.1A)

To write a percent as a decimal, drop the percent sign and multiply by 0.01. (Objective 6.1A)

To write a fraction or a decimal as a percent, multiply by 100%. (Objective 6.1B)

The basic percent equation: Part = base × rate, frequently written $P = B \times R$ (Objective 6.2A)

To translate a problem involving a percent into an equation, remember that the rate is the percent. The base usually follows the word "of." (Objective 6.2A)

Percent increase equation: Amount of increase = original value × percent increase
$$\text{Part} \quad = \quad \text{base} \quad \times \quad \text{rate}$$
$$P \quad = \quad B \quad \times \quad R$$
(Objective 6.3A)

Percent decrease equation: Amount of decrease = original value × percent decrease
$$\text{Part} \quad = \quad \text{base} \quad \times \quad \text{rate}$$
$$P \quad = \quad B \quad \times \quad R$$
(Objective 6.3B)

To solve a proportion, find the cross products. The cross products are equal. Solve the resulting equation. (Objective 6.4B)

To solve a percent problem using the proportion method, use the proportion $\frac{\text{rate}}{100} = \frac{\text{part}}{\text{base}}$. In the proportion, the rate is written as a percent without the percent sign. (Objective 6.4C)

◆ REVIEW / TEST

1. Write 75% (a) as a fraction and (b) as a decimal.

2. Write $66\frac{2}{3}$ % as a fraction.

3. Write 0.08 as a percent.

4. Write $\frac{3}{5}$ as a percent.

5. Write $\frac{7}{9}$ as a percent. Round to the nearest tenth of a percent.

6. Find 16% of 550.

7. 20 is what percent of 16?

8. 12 is 15% of what number?

9. Solve the proportion: $\frac{n}{16} = \frac{9}{4}$

10. Solve the proportion: $\frac{16}{3} = \frac{8}{n}$

11. Rosy Williamson was earning an hourly wage of $9.50 before she received an 8% raise. Find the increase in Rosy's hourly wage.

12. In 1991 Northwest Airlines' operating revenues were $7500 million. This is 82% of Delta Air Lines' operating revenues in 1991. What were Delta Air Lines' operating revenues in 1991? Round to the nearest million.

1. (a) _____

   (b) _____

2. _____

3. _____

4. _____

5. _____

6. _____

7. _____

8. _____

9. _____

10. _____

11. _____

12. _____

13. A product preference survey revealed that 9 of the 300 people surveyed liked a new product. What percent of the people surveyed did not like the product?

14. Stephanie Leone's hourly wage increased from $7.50 per hour to $8.10 per hour. Find the percent increase in Stephanie's hourly wage.

15. A newspaper reporter's salary this year is $18,000. This salary will increase by 9% next year. What will the reporter's salary be next year?

16. Next year the Montauk Company plans to decrease its $25,000 advertising budget by 2%. What will be the company's advertising budget next year?

17. Rent paid by the Berkeley Company for an office building is allocated to the company's three divisions as follows: 50%, 30%, 20%. The company pays $1275 for rent each month. Find the amount allocated to each of the three divisions.

18. Three attorneys share the profits of their firm in the ratio 3:2:1. Last year the profits were $144,000. Find the amounts received by the three partners.

19. Mandle Corporation had sales of $120,000 during its first year of operation. During the second year, sales totaled $150,000. What percent of the first year's sales were the second year's sales?

20. An investment of $9000 earns $720 each year. At the same rate, how much money must be invested to earn $1800 each year?

13. _____

14. _____

15. _____

16. _____

17. _____

    _____

    _____

    _____

18. _____

    _____

    _____

    _____

19. _____

20. _____

7

Purchasing

*O*BJECTIVES

7.1A   To calculate single trade discounts

7.1B   To calculate series trade discounts

7.1C   To convert a series discount to a single-discount equivalent

7.2A   To calculate cash discounts

7.2B   To calculate cash discounts on partial payments

PURCHASING CONSISTS OF all the activities involved in obtaining goods and services from other firms. In a small business, the owner may do the purchasing. Larger firms generally have a Purchasing Department.

The responsibilities of a Purchasing Department include locating and selecting suppliers, negotiating and placing orders, following up on the orders, receiving the goods, and making sure that deliveries are made to the right departments. The Purchasing Department must also stay abreast of current prices, quality of goods, and the reliability of suppliers—not always an easy task when there is such a large number of merchandising firms. Consider the fact that in 1990, total sales in the merchant wholesale trade in the United States were approximately $2 trillion.

The role of purchasing is critical. How well a business performs this function often makes the difference between profit and loss. For example, a small difference in price can make a big difference in cost. (Suppose you can save 3¢ a unit on a purchase of 50,000 units. What is the total amount saved?[1])

Even so, finding the right price is only one aspect of purchasing. The objective of purchasing encompasses

Purchasing the *right* goods
Having them delivered to the *right* place
In the *right* quantity
At the *right* time
For the *right* price
At the *right* quality

Why is purchasing the right quantity important? Consider a retail store. If an order for a product is too small, the store may lose customers because the product is not available for sale. If an order is too large, items will remain unsold.

Why is the right time important? To take one example, raw materials must arrive at a plant in time to be used to meet its production schedule, yet not so far ahead of time that the warehouse is overstocked with inventory. Buying at the right time may also mean buying the raw materials just before a price increase.

The major topics of this chapter are trade discounts and cash discounts, two very important considerations of purchasing.

---

[1]$1500

 **S**ECTION 7.1   **Trade Discounts**

Objective 7.1A   *To calculate single trade discounts*

Marketing is the process by which goods and services pass from one entity to another entity. This entity may be a corporation, a partnership, or an individual. Manufactured goods may pass through several entities before they reach the final consumer. A common channel of distribution is

Provider of
raw materials $\rightarrow$ manufacturer $\rightarrow$ wholesaler $\rightarrow$ retailer $\rightarrow$ consumer

Wholesalers buy finished goods in large quantities and then sell them, primarily to retailers. A wholesale distributor may have hundreds of items for sale. The distributor usually lists the suggested retail price, or **list price,** of the merchandise in a catalogue that is distributed to dealers in the company's particular trade.

The list price is not the price at which goods are sold to a retailer. Rather, list prices are the prices from which discounts are taken in order to arrive at the actual cost to the retailer. Percent discounts available to the retailer are printed on a separate sheet, called a discount sheet, and distributed with the wholesaler's catalogue. When the price of merchandise changes, only the discount sheet is adjusted; this avoids the need to reprint the catalogue.

A **trade discount** is the amount subtracted from the list price of an item and is always expressed as a percent of the list price. The basic percent equation is used to determine the discount.

Part   =   base $\times$ rate
Discount = list price $\times$ discount rate

The **net price** is the amount paid by a dealer for a product. The net price is found by subtracting the discount from the list price.

Net price = list price – discount

A manufacturer's catalogue lists the price of a calculator as $79.50. A wholesaler can buy 1000 of these calculators at a discount of 25% off the list price. Find the discount and the net price of the 1000 calculators.

♦ Find the total list price by multiplying the list price by the number of items purchased.

$79.50(1000) = 79,500$

♦ Find the discount.

Discount = list price $\times$ discount rate
$= (79,500)(0.25)$
$= 19,875$

The discount is $19,875.

♦ The net price is the list price minus the discount.

Net price = list price – discount
$= 79,500 - 19,875$
$= 59,625$

The net price is $59,625.

An alternative method of finding the net price is used when it is not necessary to know the discount. This method involves use of the complement of the discount rate. The sum of a percent discount and its complement is 100%. Thus the complement of a 25% discount is 100% − 25% = 75%. The complement is the percent of the list price that the customer pays for the item. For example, if a 25% discount is allowed for an item, then 75% of the list price is the amount the customer pays for the item.

Net price = list price × the complement of the discount rate

Find the net price of a tire listed at $80 that is selling at a 40% discount.

* Multiply the list price by the complement of the discount rate (100% − 40% = 60%).

Net price = 80(0.60)
                 = 48

The net price is $48.

One wholesaler lists a personal computer at $2750 with a trade discount of 25%. Another wholesaler lists the same computer at $2250 with a trade discount of 15%. Which is the better buy?

* Multiply the list price of the first computer by the complement of the discount rate (100% − 25% = 75%).

Net price = (2750)(0.75)
                 = 2062.50

The net price of the first computer is $2062.50.

* Multiply the list price of the second computer by the complement of the discount rate (100% − 15% = 85%).

Net price = (2250)(0.85)
                 = 1912.50

The net price of the second computer is $1912.50.

* Compare the two prices.

$1912.50 is less than $2062.50.

The computer with a list price of $2250 and a 15% discount is the better buy.

When the net price and the trade discount rate are known, the list price can be found by dividing the net price by the complement of the trade discount rate.

$$\text{List price} = \frac{\text{net price}}{\text{complement of the trade discount rate}}$$

The net price of a washing machine with a 30% trade discount is $350. Find the list price.

* Divide the net price by the complement of the trade discount rate (100% − 30% = 70%).

$$\text{List price} = \frac{350}{0.70}$$
$$= 500$$

The list price is $500.

Example 1

A lathe is listed at $7200 with a trade discount of 30%. Find the discount and the net price.

Strategy

* To find the discount, multiply the discount rate (30%) by the list price (7200).

* To find the net price, subtract the discount from the list price.

Solution

Discount = list price × discount rate
$$= 7200(0.30)$$
$$= 2160$$

The discount is $2160.

Net price = list price − discount
$$= 7200 − 2160$$
$$= 5040$$

The net price is $5040.

You Try It 1

A shipment of radios is listed at $32,500 with a trade discount of 20%. Find the discount and the net price.

Your strategy

Your solution

Example 2

The list price of a milling machine is $12,500. A trade discount of 22% applies. Find the net price.

Strategy

To find the net price, multiply the list price by the complement of the discount rate (100% − 22% = 78%).

Solution

Net price = 12,500(0.78)
$$= 9750$$

The net price is $9750.

You Try It 2

A lawn tractor with a list price of $1750 is advertised at a trade discount of 45%. Find the net price.

Your strategy

Your solution

Solutions on p. A23

Example 3

The net price of a fax machine with a 40% trade discount is $360. Find the list price.

Strategy

To find the list price, divide the net price (360) by the complement of the trade discount rate (100% − 40% = 60%).

Solution

$$\text{List price} = \frac{360}{0.60}$$

$$= 600$$

The list price is $600.

You Try It 3

Find the list price of a cellular phone with a net price of $225 and a trade discount of 25%.

Your strategy

Your solution

Solution on p. A23

Objective 7.1B *To calculate series trade discounts*

A manufacturer or wholesaler may offer one or more discounts, depending on the buyer and on market conditions. Three examples are given below.

1. A manufacturer sells products to a retailer at a 25% discount.

2. A manufacturer sells a small order to a wholesaler and gives a discount of 10% in addition to the 25% discount given to the retailer. The two discounts would be represented as 25/10.

3. A manufacturer sells a large order to a wholesaler and gives a 5% discount in addition to the 25% and the 10% discounts. The three discounts would be represented as 25/10/5.

Two or more discounts given on the same item constitute a **series discount,** or **chain discount.** In a series discount, the discount rates cannot be added. The discounts are subtracted successively from the list price. In a 25/10 discount, 25% of the list price is deducted from the list price to find the first net price. Then 10% of the first net price is deducted to find the final cost to the buyer. This process is illustrated in the following example.

A wholesaler offers a series discount of 25/10/5 off the list price of $200,000 on an order of bicycles. Find the final cost (1) by using the discounts and (2) by using the complements of the discount rates.

(1) By using discounts:

* Subtract the first discount $(200{,}000 \times 0.25)$ from the list price.

$$\begin{array}{r} 200{,}000 \\ -\ 50{,}000 \end{array}$$

* Subtract the second discount $(150{,}000 \times 0.10)$ from the net price after the first discount.

$$\begin{array}{r} 150{,}000 \\ -\ 15{,}000 \end{array}$$

* Subtract the third discount $(135{,}000 \times 0.05)$ from the net price after the second discount.

$$\begin{array}{r} 135{,}000 \\ -\ \ 6{,}750 \\ \hline 128{,}250 \end{array}$$

The net price is $128,250.

(2) By using complements:

* Multiply the list price by the complement of 25%.

$$\begin{array}{r} 200{,}000 \\ \times\ \ \ \ 0.75 \end{array}$$

* Multiply 150,000 by the complement of 10%.

$$\begin{array}{r} 150{,}000 \\ \times\ \ \ \ 0.90 \end{array}$$

* Multiply 135,000 by the complement of 5%.

$$\begin{array}{r} 135{,}000 \\ \times\ \ \ \ 0.95 \\ \hline 128{,}250 \end{array}$$

The net price is $128,250.

In this example, the net price of a $200,000 order with a series discount of 25/10/5 is $128,250. The same result could have been obtained by multiplying the list price by the product of the complements of the discount rates.

> **Net price = list price $\times$ product of the complements of the discount rates**
> $$\begin{aligned} &= 200{,}000 &&\times (0.75)(0.90)(0.95) \\ &= 200{,}000 &&\times (0.64125) \\ &= 128{,}250 \end{aligned}$$

The product of the complements of the discount rates is 0.64125. Converting this decimal to a percent (64.125%) means that the net price is 64.125% of the list price.

Note that the decimal equivalent 0.64125 was not rounded before it was multiplied by the list price. Although a net price is always rounded to the nearest cent, the product of the complements of the discount rates should not be rounded.

The series discount 25/10/5 was used in this example. The same net price would have been obtained if the discount rates were in a different order, such as 10/25/5 or 5/10/25. However, the largest discount rate is usually listed first.

Example 4

A microcomputer system is available to a retailer for a price of $3200 with a series discount of 30/15. Find the second discount.

Strategy

To find the second discount:

* Multiply the list price by the first discount rate (30%) to find the first discount.

* Subtract the first discount from the list price to find the first net price.

* Multiply the first net price by the second discount rate (15%) to find the second discount.

Solution

First discount = 3200(0.30) = 960

First net price = 3200 − 960 = 2240

Second discount = 2240(0.15) = 336

The second discount is $336.

You Try It 4

Hudson's Hallmark Shop orders $2500 of merchandise and receives a series discount of 15/10. Find the second discount.

Your strategy

Your solution

Example 5

Find the net price of a portable computer with a list price of $1200 and a series discount of 30/20/10.

Strategy

To find the net price, multiply the list price by the product of the complements of the discount rates.

Solution

Net price = 1200 × (0.70)(0.80)(0.90)
 = 1200 × (0.504) = 604.80

The net price is $604.80.

You Try It 5

A True Value Hardware Store orders $18,000 of merchandise and receives a series discount of 15/10/5. Find the net price.

Your strategy

Your solution

Solutions on p. A23

Objective 7.1C *To convert a series discount to a single-discount equivalent*

When the same series discount is used repeatedly, it is convenient to have a single discount rate that yields the same discount as the series of discounts. A single discount rate that is equivalent to a series discount is called a **single-discount equivalent.** Using it can greatly reduce the number of computations needed to determine net prices.

Find the single-discount equivalent that would yield the same net price for an order with a list price of $1000 less a series discount of 20/10.

* Find the net price by multiplying the list price by the complements of the discount rates.

$$\text{Net price} = 1000 \times (0.80)(0.90)$$
$$= 1000 \times (0.72)$$
$$= 720$$

* Find the discount.

$$\text{Discount} = \text{list price} - \text{net price}$$
$$= 1000 - 720$$
$$= 280$$

* Use the basic percent equation to find the single-discount equivalent.

$$\text{Part} = \text{base} \times \text{rate}$$
$$280 = 1000r$$
$$\frac{280}{1000} = r$$
$$0.28 = r$$

The single-discount equivalent for a series discount of 20/10 is 28%.

Note that the single-discount equivalent (28%) is the complement of the product of the complements of the discount rates (0.72 = 72%).

$$\textbf{Single-discount equivalent} = \textbf{100\%} - \begin{array}{l} \textbf{product of the complements} \\ \textbf{of the discount rates} \end{array}$$

Find the single-discount equivalent of a 25/15/5 series discount.

* Subtract the product of the complements of the discount rates from 1.00.

$$1.00 - (0.75)(0.85)(0.95) =$$
$$1.00 - (0.605625) =$$
$$0.394375$$

The single-discount equivalent of a 25/15/5 series discount is 39.4375%.

Finding the single-discount equivalents of series discounts can be used to determine which of two purchases is the better buy.

Two different manufacturers are offering a comparable item at the same list price, the first with a discount of 30/10 and the second with a discount of 25/15. Which is the better buy?

* Find the single-discount equivalent of a 30/10 series discount.

$$1.00 - (0.70)(0.90) = 1.00 - 0.63$$
$$= 0.37$$
$$= 37\%$$

* Find the single-discount equivalent of a 25/15 series discount.

$$1.00 - (0.75)(0.85) = 1.00 - 0.6375$$
$$= 0.3625$$
$$= 36.25\%$$

* A discount rate of 37% is greater than a discount rate of 36.25%.

The first manufacturer is offering the better buy.

Example 6

Find the single-discount equivalent of a 15/10/5 series discount.

Strategy

To find the single-discount equivalent, subtract the product of the complements of the discount rates from 1.00.

Solution

$1.00 - (0.85)(0.90)(0.95) = 1.00 - 0.72675$
$= 0.27325$

The single-discount equivalent is 27.325%.

You Try It 6

Find the single-discount equivalent of an 18/10/4 series discount.

Your strategy

Your solution

Example 7

A Lion Food Store orders soft drinks listed at $8000 and receives a series discount of 25/10. Find the discount.

Strategy

To find the discount:

* Find the single-discount equivalent of a 25/10 series discount.

* Multiply the list price by the single-discount equivalent.

Solution

$1.00 - (0.75)(0.90) = 1.00 - 0.675$
$= 0.325$

Discount $= 8000(0.325) = 2600$

The discount is $2600.

You Try It 7

A Goodyear Tire dealer orders new tires listed at $12,000 and receives a series discount of 20/5. Find the discount.

Your strategy

Your solution

Solutions on pp. A23–A24

EXERCISE 7.1A

Solve.

1. A wholesale shipment of greeting cards with a catalogue list price of $5000 is sold at a discount rate of 30%. Find (a) the discount and (b) the net price to the retailer.

2. A wholesale distributor of gold jewelry is offering the entire stock for 60% off the catalogue price. For a gold chain costing $1500, find (a) the trade discount and (b) the net price.

3. Find the complement of a 45% discount rate. Express the complement (a) as a percent and (b) as a decimal.

4. Find the complement of a 34% discount rate. Express the complement (a) as a percent and (b) as a decimal.

5. A washing machine with a list price of $315 is offered to a retail store at 25% off the catalogue price. Find (a) the complement of the discount rate and (b) the net price of the washing machine.

6. A lathe with a list price of $8200 is advertised at a trade discount rate of 35%. Find (a) the complement of the discount rate and (b) the net price of the lathe.

7. A word processor is listed at $2200 with a trade discount rate of 45%. Find (a) the discount and (b) the net price.

8. A shipment of paint is listed at $7500 with a trade discount of 40%. Find (a) the discount and (b) the net price.

9. A manufacturer's catalogue lists the price of an answering machine as $80. A wholesaler can purchase 500 of the answering machines with a 25% trade discount. Find the net price of the 500 answering machines.

10. A manufacturer's catalogue lists the price of a portable CD player as $120. A wholesaler purchases 1000 of the CD players with a 40% trade discount. Find the net price of the 1000 CD players.

11. A wholesaler can purchase a cordless telephone with a 20% trade discount. The list price is $90. Find the net price if the wholesaler buys 250 cordless telephones.

12. The list price of a camcorder is $800. Find the net price of 100 camcorders if a wholesaler receives a 30% trade discount.

1. (a) _____

   (b) _____

2. (a) _____

   (b) _____

3. (a) _____

   (b) _____

4. (a) _____

   (b) _____

5. (a) _____

   (b) _____

6. (a) _____

   (b) _____

7. (a) _____

   (b) _____

8. (a) _____

   (b) _____

9. _____

10. _____

11. _____

12. _____

13. One wholesaler lists a calculator at $40 with a discount rate of 35%. Another wholesaler lists the same calculator at $30 with a discount rate of 20%. Which is the better buy?

14. One wholesaler lists a speaker phone at $60 with a discount rate of 20%. Another wholesaler lists the same speaker phone at $55 with a discount rate of 25%. Which is the better buy?

15. The net price of a sewing machine with a 25% trade discount is $300. Find the list price.

16. The net price of a trash compactor with a 35% trade discount is $227.50. Find the list price.

17. A wholesaler pays a net price of $570 on a 5-horsepower motor with a 40% trade discount. What is the list price?

18. A wholesaler pays a net price of $105 on a dome tent with a 30% trade discount. What is the list price?

Exercise 7.1B

Solve.

19. A laser printer is available to a retailer for a price of $2200 with a series discount of 30/10. Find (a) the discount and (b) the net price.

20. A set of golf clubs is purchased by a retailer for a price of $200 with a series discount of 25/5. Find (a) the discount and (b) the net price.

21. The list price of a pool table is $1200 with a series discount of 25/10. What is the amount of the second trade discount?

22. The list price of a motor boat is $4200 with a series discount of 30/10. What is the amount of the second trade discount?

23. Find the decimal equivalent of the net price for a 30/10/5 series trade discount.

24. Find the decimal equivalent of the net price for a 40/20/10 series trade discount.

25. A May Department Store orders $22,000 of merchandise and receives a series discount of 30/15. Find the net price of the order.

26. A Toys "Я" Us store orders $7000 of merchandise and receives a series discount of 25/10. Find the net price of the order.

13. _____

14. _____

15. _____

16. _____

17. _____

18. _____

19. (a) _____

    (b) _____

20. (a) _____

    (b) _____

21. _____

22. _____

23. _____

24. _____

25. _____

26. _____

27. An automotive parts wholesaler offers a series discount of 25/15. Find the net price of an item listed for $920.

27. _____

28. A home supply wholesaler offers a series discount of 20/10. Find the net price of a recliner listed for $240.

28. _____

29. A manufacturer's list price for a suit is $225 with a series discount of 40/20/10. Find the net price.

29. _____

30. _____

30. Find the net price of a coat with a list price of $290 and a series discount of 30/20/10.

31. _____

32. _____

31. A Winn-Dixie store orders $9000 of merchandise and receives a series discount of 20/10/5. Find the net price.

33. _____

34. _____

32. A Woolworth store orders $14,000 of merchandise and receives a series discount of 25/15/5. Find the net price.

35. (a) _____

(b) _____

33. A gym orders $75,000 of equipment and receives a series discount of 25/20/10. Find the net price.

36. (a) _____

(b) _____

34. An accountant orders $30,000 of office equipment and receives a series discount of 20/12/8. Find the net price.

37. _____

38. _____

EXERCISE 7.1C

Solve.

35. Find the single-discount equivalent of a 40/10/5 series discount. Express the discount (a) as a percent and (b) as a decimal.

36. Find the single-discount equivalent of a 30/20/10 series discount. Express the discount (a) as a percent and (b) as a decimal.

37. A Payless Shoe Store orders $4000 of shoes and receives a series discount of 30/5. Find the discount.

38. The Ashland Exxon Service Station orders $6000 of tires and receives a series discount of 40/15. Find the discount.

39. A clothing manufacturer offers a series discount of 30/10/5 on all goods. Find the discount on an order of $10,000.

40. A spa is listed for $2250 with a series discount of 15/10/5. Find (a) the discount and (b) the net price.

41. A tractor is listed for $3200 with a series discount of 20/10/5. Find (a) the discount and (b) the net price.

42. A shipment of hardware is listed for $5500 with a series discount of 20/12/5. Find (a) the discount and (b) the net price.

43. A shipment of shirts is listed for $3000 with a series discount of 25/12/5. Find (a) the discount and (b) the net price.

44. The list price of a camera is $345. A series discount of 20/10/5 is offered on all orders greater than $3000. Find the discount for an order of 15 cameras.

45. The list price of a VCR is $460. A series discount of 15/10/5 is offered on all orders greater than $5000. Find the discount for an order of 20 VCRs.

46. The list price of a suit is $240. A series discount of 25/15/5 is offered on all orders greater than $10,000. Find (a) the discount and (b) the net price on an order of 50 suits.

47. Two different manufacturers are offering comparable items at the same list price, the first with a series discount of 30/20/5 and the second with a series discount of 40/10/5. Which is the better buy?

48. Two different manufacturers are offering comparable items at the same list price, the first with a series discount of 25/15/10 and the second with a series discount of 30/15/5. Which is the better buy?

49. The list price of a computer system is $2450 with a series discount of 30/10/5. Another distributor lists the same system at a price of $2000 and a series discount of 25/10. Which is the better buy?

50. The list price of a washer-dryer is $720 with a series discount of 30/20/5. Another distributor lists the same washer-dryer at a price of $850 and a series discount of 40/20. Which is the better buy?

39. _____

40. (a) _____

(b) _____

41. (a) _____

(b) _____

42. (a) _____

(b) _____

43. (a) _____

(b) _____

44. _____

45. _____

46. (a) _____

(b) _____

47. _____

48. _____

49. _____

50. _____

Cash Discounts

Objective 7.2A   *To calculate cash discounts*

Manufacturers and wholesalers often borrow money to pay for raw materials or products and to operate their businesses. To reduce the need for borrowing, a business may offer a cash discount to customers. **Cash discounts** are reductions in the net price of a product sold on credit and are given to encourage the prompt payment of a bill, or **invoice.** The cash discount is usually a percent of the net, or invoice, price.

Cash discount = net price × percent discount

The amount due is the net price minus the cash discount, or the product of the net price and the complement of the cash discount rate.

Amount due = net price − cash discount
Amount due = net price × complement of the cash discount rate

To receive a cash discount, a customer must pay the invoice within a specified period of time. Some common cash discount terms and their meanings are listed in the following table.

Term	Meaning
n/30	The full amount is due within 30 days of the date of the invoice.
n/EOM	The full amount is due at the end of the month.
n/10 EOM	The full amount is due 10 days after the end of the month.
n/10 ROG	The full amount is due 10 days after receipt of the goods.
2/10, n/30	A 2% discount is given if payment is made within 10 days after the date of the invoice. The full amount is due 30 days after the date of the invoice.
2/10, n/30 EOM	A 2% discount is given if payment is made within 10 days after the end of the month in which the invoice is dated. The full amount is due 30 days after the end of the month.
2/10, n/30 ROG	A 2% discount is given if payment is made 10 days after receipt of the goods. The full amount is due 30 days after receipt of the goods.

In this list of terms, n represents "net," EOM represents "end of month," and ROG represents "receipt of goods." 2/10, n/30 is read "two ten, net thirty"; the first number is the cash discount rate (2%), the second number is the number of days in which the cash discount may be taken, and 30 is the number of days the customer has to pay the invoice in full. After 30 days, the invoice is overdue and may be subject to late charges.

To find the due date of an invoice, use the number of days in each month.

30-day months:	April, June, September, November
31-day months:	January, March, May, July, August, October, December
28-day month:	February
Exception:	February has 29 days in a leap year.

Leap years are divisible by 4. For example, 1996 is divisible by 4 and 1996 is therefore a leap year.

Find the date 10 days from May 8.

* Add 10 to the day of the month. $8 + 10 = 18$

10 days from May 8 is May 18.

Find the date 10 days from March 25.

* There are fewer than 10 days left in March.
  There are 6 days from March 25 to March 31. $31 - 25 = 6$

* To have a total of 10 days, 4 days in April are needed. $10 - 6 = 4$

10 days from March 25 is April 4.

An invoice for $7400 is dated July 27, and terms of 2/10, n/30 are offered. Find the amount due if the invoice is paid within 10 days.

* Multiply the net price by the complement   Amount due $= 7400(0.98)$
  of the cash discount rate of 2%. $= 7252$

The amount due is $7252.

An invoice is a printed record of the transaction between buyer and seller. It contains all information pertinent to the transaction. An example of an invoice follows. Note that the cash discount period starts with the date of the invoice and that both trade and cash discounts are offered.

PURCHASE ORDER NO	SHIPPED VIA	TERMS	DATE SHIPPED
10276	Fr.	2/10, n/30	4/7

Quantity	Description		Unit Price	Amount
12	Steel radial P 165/80 R-13		$24.00	
6	Steel radial P 185/75 R-14		$25.50	
24	Steel radial P 205/75 R-14		$26.00	
12	Steel radial P 215/75 R-15		$28.00	
24	Steel radial P 225/75 R-15		$31.50	
		Total List Price		
		Less Trade Discounts of 15/5		
		Amount Due		
		Less Cash Discount If Paid by (4/17)		
		Total Amount Paid		

Some figures are not provided in the above invoice. Using the information provided, calculate the total amount paid if payment is made by April 17.

- Find the total list price by multiplying the unit price by the quantity for each item listed and then adding the products.

$$
\begin{array}{rcl}
12 \times 24.00 & = & 288.00 \\
6 \times 25.50 & = & 153.00 \\
24 \times 26.00 & = & 624.00 \\
12 \times 28.00 & = & 336.00 \\
24 \times 31.50 & = & 756.00 \\
\hline
\text{Total list price} & = & 2157.00
\end{array}
$$

- Find the net price, or the amount due after the trade discounts, by multiplying the total list price by the product of the complements of the trade discount rates.

$$
\begin{aligned}
\text{Net price} &= 2157.00 \times (0.85)(0.95) \\
&\approx 1741.78
\end{aligned}
$$

- Multiply the net price by the complement of the cash discount rate.

$$
\begin{aligned}
\text{Amount paid} &= 1741.78(0.98) \\
&\approx 1706.94
\end{aligned}
$$

The amount paid is $1706.94.

Example 1

An invoice for $24,000 is dated April 12, and terms of 3/10, n/30 are offered. Find the due date for the cash discount and the amount due if the invoice is paid within 10 days.

Strategy

- To find the due date, add 10 to the day of the month.

- To find the amount due, multiply the net price by the complement of the cash discount rate.

Solution

$12 + 10 = 22$

The due date is April 22.

Amount due $= 24,000(0.97) = 23,280$

If the invoice is paid by April 22, the amount due is $23,280.

You Try It 1

An invoice for $3700 is dated November 24. The terms are 4/15, n/60. Find the due date for the cash discount and the amount due if the invoice is paid within 15 days.

Your strategy

Your solution

Solution on p. A24

Example 2

An invoice for $12,500 is dated January 28. The terms are 3/15, n/30. Find the cash discount if the invoice is paid within 15 days.

Strategy

To find the cash discount, multiply the amount due by the cash discount rate.

Solution

Cash discount = 12,500(0.03) = 375

The cash discount is $375.

Example 3

Find the amount due if this invoice is paid on November 22 with a trade discount of 20/5.

TERMS 3/10, n/30	DATE 11/15	
QUANTITY	UNIT PRICE	EXTENSION
40	$5.15	
75	$13.50	

Strategy

To find the amount due:

- Find the total list price by calculating the sum of the extensions.

- Find the net price by multiplying the total list price by the product of the complements of the trade discounts.

- Multiply the net price by the complement of the cash discount rate of 3%.

Solution

$$
\begin{aligned}
40 \times 5.15 &= 206.00 \\
75 \times 13.50 &= \underline{1012.50} \\
\text{Total list price} &= 1218.50
\end{aligned}
$$

$$
\begin{aligned}
\text{Net price} &= 1218.50 \times (0.80)(0.95) \\
&= 926.06
\end{aligned}
$$

$$
\begin{aligned}
\text{Amount due} &= 926.06 \times 0.97 \\
&\approx 898.28
\end{aligned}
$$

The amount due is $898.28.

You Try It 2

An invoice for $8750 is dated September 10. The terms are 3/10, n/30. Find the cash discount if the invoice is paid within 10 days.

Your strategy

Your solution

You Try It 3

Find the amount due if this invoice is paid on March 18 with a trade discount of 15/10.

TERMS 5/10, n/30	DATE 3/12	
QUANTITY	UNIT PRICE	EXTENSION
80	$10.55	
25	$85.40	

Your strategy

Your solution

Solutions on p. A 24–A 25

Objective 7.2B   *To calculate cash discounts on partial payments*

A customer may not have enough cash to pay the full amount of an invoice but may have enough funds to pay a portion of it. In that case, the customer may pay part of the full amount within the discount period and be credited with the discount for that portion paid.

An invoice for $4500 is dated July 10, and terms of 3/10, n/30 are offered. A partial payment of $2000 is made on July 18. Find the amount credited to the bill and the amount still due on the invoice after that payment is made.

◆ Because the partial payment is made within 10 days, the 3% discount applies. To find the amount credited, use the basic percent equation. The part is the payment, the base is the amount credited, and the rate is the complement of the cash discount rate. ($2000 is 97% of the amount credited to the bill.)

$$\text{Part} = \text{base} \times \text{rate}$$
$$\text{Payment} = \text{amount credited} \times \text{rate}$$
$$2000 = \text{amount credited} \times 0.97$$
$$\frac{2000}{0.97} = \text{amount credited}$$
$$2061.86 \approx \text{amount credited}$$

The amount credited to the bill is $2061.86.

◆ To find the amount due, subtract the amount credited from the amount of the invoice.

$$\text{Amount due} = \text{amount of invoice} - \text{amount credited}$$
$$= 4500 - 2061.86$$
$$= 2438.14$$

The amount due on the bill after the payment is made is $2438.14.

Example 4

An invoice of $15,000 is dated August 8, and terms of 3/10, n/30 are offered. A partial payment of $5000 is made on August 16. Find the amount credited to the bill.

Strategy

To find the amount credited, use the basic percent equation. The rate is the complement of the cash discount rate.

Solution

$$\text{Part} = \text{base} \times \text{rate}$$
$$\text{Payment} = \text{amount credited} \times \text{rate}$$
$$5000 = \text{amount credited} \times 0.97$$
$$\frac{5000}{0.97} = \text{amounted credited}$$
$$5154.64 \approx \text{amount credited}$$

The amount credited to the bill is $5154.64.

You Try It 4

An invoice for $23,000 is dated June 16, and terms of 2/10, n/30 are offered. A partial payment of $10,000 is made on June 24. Find the amount credited to the bill.

Your strategy

Your solution

Solution on p. A25

Example 5

A lumber mill offers a cash discount of 5/10, n/30 on an order of $62,000. The invoice is dated September 14. A payment of $30,000 is made on September 23, and the remainder is paid on October 10. Find the total amount paid.

You Try It 5

A wholesaler of china offers cash discounts with terms 4/10, n/30. An invoice for $3200 is dated July 8. A partial payment of $1200 is made on July 16, and the remainder is paid on August 6. Find the total amount paid.

Strategy

To find the total amount paid:

- Find the amount credited for the first payment.

- Find the amount due by subtracting the amount credited from the amount of the invoice.

- Add the first payment to the amount due.

Your strategy

Solution

$$\text{Part} = \text{base} \times \text{rate}$$

$$\text{Payment} = \text{amount credited} \times \text{rate}$$

$$30,000 = \text{amount credited} \times 0.95$$

$$\frac{30,000}{0.95} = \text{amount credited}$$

$$31,578.95 \approx \text{amount credited}$$

$$\text{Amount due} = \frac{\text{amount}}{\text{of invoice}} - \frac{\text{amount}}{\text{credited}}$$

$$= 62,000 - 31,578.95$$

$$= 30,421.05$$

$$\text{Total paid} = 30,000 + 30,421.05$$

$$= 60,421.05$$

The total amount paid was $60,421.05.

Your solution

Solution on p. A25

*E*XERCISE 7.2A

Solve.

1. An invoice of $4500 is dated May 6, and terms of 4/10, n/30 are offered. Find the discount if the invoice is paid within 10 days.

2. An invoice of $1750 is dated July 12, and terms of 3/10, n/30 are offered. Find the discount if the invoice is paid within 10 days.

3. A Mobil service station receives an invoice of $755 dated July 28, and terms of 5/10, n/30 are offered. Find the amount due if the invoice is paid within 10 days.

4. A Wal-Mart store receives an invoice of $680 dated March 25, and terms of 4/10, n/30 are offered. Find the amount due if the invoice is paid within 10 days.

5. An invoice of $3000 is dated June 26, and terms of 3/10, n/30 are offered. Find (a) the cash discount and (b) the amount due if the invoice is paid within 10 days.

6. An invoice of $1860 is dated April 22, and terms of 4/10, n/30 are offered. Find (a) the cash discount and (b) the amount due if the invoice is paid within 10 days.

7. A Motorola dealer receives an invoice of $5200 dated March 17, and terms of 4/10, n/30 EOM are offered. Find the amount due if the discount is earned.

8. A Safeway retail store receives an invoice of $2150 dated July 12, and terms of 3/10, n/30 EOM are offered. Find the amount due if the discount is earned.

9. An invoice of $11,420 is dated July 28, and terms of 5/10, n/30 EOM are offered. Find (a) the last date the discount may be taken, (b) the due date of the invoice, and (c) the amount due if the discount is earned.

1. _____

2. _____

3. _____

4. _____

5. (a) _____

   (b) _____

6. (a) _____

   (b) _____

7. _____

8. _____

9. (a) _____

   (b) _____

   (c) _____

10. An invoice of $2150 is dated July 12, and terms of 3/10, n/30 EOM are offered. Find (a) the last day the discount may be taken, (b) the due date of the invoice, and (c) the amount due if the discount is earned.

11. A Publix Super Market receives an invoice of $2245 dated May 27, and terms of 2/10, n/30 are offered. Find (a) the last day the discount may be taken, (b) the due date of the invoice, and (c) the amount due if the discount is earned.

12. A Texaco service station receives an invoice of $866 dated March 24, and terms of 4/10, n/30 are offered. Find (a) the last day the discount may be taken, (b) the due date of the invoice, and (c) the amount due if the discount is earned.

The following invoice is dated May 2. The invoice is paid on May 10. Calculate the extensions, the total list price, the series discount, the net price, the cash discount, and the amount paid. The series discount offered is 20/5, and the cash terms are 5/10, n/30.

Quantity	Description	Unit Price	Amount
20	Trench coat	48.20	13.
50	Long-sleeved shirt	6.70	14.
25	Suits	106.75	15.
24	Shoes	23.40	16.
48	Travel bag	5.80	17.
		Total List Price	18.
		Series Discount	19.
		Net Price	20.
		Cash Discount	21.
		Total Amount Paid	22.

10. (a) _____

(b) _____

(c) _____

11. (a) _____

(b) _____

(c) _____

12. (a) _____

(b) _____

(c) _____

13. _____

14. _____

15. _____

16. _____

17. _____

18. _____

19. _____

20. _____

21. _____

22. _____

The following invoice, dated April 25, offers a series discount of 15/10 and a cash discount of 3/10, n/30. The invoice is paid on May 3. Calculate the extensions, the total list price, the series discount, the net price, the cash discount, and the amount to be paid.

Quantity	Description	Unit Price	Amount
5	Video Camera	914.50	23.
3	Wide angle lens	74.60	24.
3	Telephoto lens	108.40	25.
64	8 mm video tape	9.25	26.
12	VCR	419.25	27.
		Total List Price	28.
		Trade Discount	29.
		Net Price	30.
		Cash Discount	31.
		Total Amount Paid	32.

The following invoice is dated August 15. The series discount offered is 20/10, and the cash discount terms are 4/10, n/30 EOM. The invoice is paid on September 8. Calculate the extensions, the total list price, the series discount, the net price, the cash discount, and the amount paid.

Quantity	Description	Unit Price	Amount
24	Tents	174.50	33.
36	Propane stoves	32.25	34.
36	Lantern	11.30	35.
30	Sleeping bag	39.70	36.
12	Canopy	88.40	37.
		Total List Price	38.
		Trade Discount	39.
		Net Price	40.
		Cash Discount	41.
		Total Amount Paid	42.

23. _____
24. _____
25. _____
26. _____
27. _____
28. _____
29. _____
30. _____
31. _____
32. _____
33. _____
34. _____
35. _____
36. _____
37. _____
38. _____
39. _____
40. _____
41. _____
42. _____

Exercise 7.2B

Solve.

43. An invoice for $6400 is dated May 5, and terms of 4/10, n/30 are offered. A partial payment of $3000 is made on May 14. Find the amount credited to the bill.

44. An invoice for $1650 is dated April 30, and terms of 2/10, n/30 are offered. A partial payment of $800 is made on May 8. Find the amount credited to the bill.

45. A Walgreen's store receives an invoice for $1850 dated June 12, and terms of 5/10, n/30 are offered. A partial payment of $1000 is made on June 20. Find the amount credited to the bill.

46. One of R. H. Macy's department stores receives an invoice for $1130 dated August 15, and terms of 3/10, n/30 are offered. A partial payment of $500 is made on August 22. Find the amount credited to the bill.

47. A contractor receives an order of construction materials costing $32,000 with terms of 4/10, n/30. The invoice is dated May 5, and a $20,000 payment is made on May 14. Find the amount remaining to be paid.

48. A plumber receives an order of plumbing materials costing $18,500 with terms of 2/10, n/30. The invoice is dated March 10, and a $10,000 payment is made on March 19. Find the amount remaining to be paid.

49. An invoice for $3000 is dated April 14, and terms of 4/10, n/30 are offered. A payment of $2000 is made on April 20, and the remainder is paid on May 12. Find the total amount paid.

50. An invoice for $1400 is dated May 16, and terms of 3/10, n/30 are offered. A payment of $1000 is made on May 25, and the remainder is paid on June 10. Find the total amount paid.

43. _____

44. _____

45. _____

46. _____

47. _____

48. _____

49. _____

50. _____

CALCULATORS

Store/Recall Keys   In the following table, the list price and discount rate on an item are given, and both the amount of discount and the net price are to be calculated. Enter the list price in your calculator, and store it in the calculator's memory. Multiply the list price by the discount rate to determine the amount of discount. Recall the list price from the calculator's memory, and multiply it by the complement of the discount rate to determine the net price. Consult the operator's manual for your calculator to determine what key stores a number in memory and what key recalls a number from memory.

Complete the following table. Round to the nearest cent.

List Price	Discount Rate	Amount of Discount	Net Price
$431.98	5%	1.	2.
$584.32	10%	3.	4.
$647.95	25%	5.	6.
$3575.40	25%	7.	8.
$1890.75	20%	9.	10.
$2843.39	10.5%	11.	12.
$3742.99	12.5%	13.	14.

Complete the following table. Round to the nearest cent.

Net Price	Cash Discount Rate	Amount of Discount	Total Amount Due
$672.90	2%	15.	16.
$1453.50	2%	17.	18.
$2798.35	1%	19.	20.
$5684.20	3%	21.	22.

1. _____
2. _____
3. _____
4. _____
5. _____
6. _____
7. _____
8. _____
9. _____
10. _____
11. _____
12. _____
13. _____
14. _____
15. _____
16. _____
17. _____
18. _____
19. _____
20. _____
21. _____
22. _____

BUSINESS CASE STUDY

Choosing Suppliers   You work in the Purchasing Department of a large department store and are responsible for the purchases for the infants' department. On October 24, you received a shipment from the Concord Company, a wholesaler. The shipment was purchased with a series discount of 20/5.

A portion of the invoice that arrived with the shipment is shown here. The cash terms of the invoice are 2/10, 1/15, n/30, which means that a 2% discount can be taken within 10 days of the invoice, and that if the company does not take advantage of the 2% discount, a 1% discount can be taken within 11 to 15 days of the invoice; the full amount is to be paid within 30 days of the date of the invoice.

Quantity	Catalogue Number	Description	Unit Price	Amount
50	BP-8033	Bumper pad	25.00	_____
25	HC-3729	Maple highchair	55.00	_____
20	MP-5216	Mesh playpen	60.00	_____
15	PC-2473	Porta-crib	90.00	_____
20	RG-7904	Restraining gate	20.00	_____
50	US-4651	Umbrella stroller	50.00	_____
			Total List Price	_____
			Trade Discount	_____
			Net Price	_____
			Transportation	175.00
			Cash Discount	_____
			Total Amount Paid	_____

Upon inspecting the shipment, you discovered that 3 of the bumper pads, 1 of the mesh playpens, and 2 of the umbrella strollers are defective. You have arranged to have these items returned to the wholesaler. The company will not be charged for these items, and the invoice must be adjusted accordingly.

This is not the first time that a shipment from this supplier has contained a number of defective items. You are aware that there is a cost to your company when goods must be returned: it is the company's money that pays its employees for the time they spend buying merchandise, receiving and inspecting purchases, and arranging for returns.

You have done some business with another supplier, the Lexington Company, which sells merchandise comparable to that sold by the Concord Company. Both the Concord Company and the Lexington Company deliver goods on time and fill orders accurately. Generally, the Lexington Company offers you a series discount of 15/5 and cash terms of 1/10, n/30.

Upon checking your records, you find that for all shipments received from the Lexington Company, the percent of defective or damaged goods is 1%. You consider this an acceptable level.

1. Calculate the total amount due the Concord Company if the invoice is paid on November 2. Remember that the defective items are not paid for by the company. Also, transportation charges are not subject to the cash discount.

2. Calculate the total amount due the Concord Company if the invoice is paid on November 6.

3. What is the amount due the Concord Company if the invoice is paid on November 20?

4. What is the latest date by which the invoice must be paid?

5. Assume that one partial payment of $4000 is made on November 1 and that a second partial payment of $2000 is made on November 5. Find the amount that remains to be paid by the date 30 days from the date of the invoice.

6. What percent of the items shipped by the Concord Company were defective? How does this compare with the percent of damaged goods received from the Lexington Company?

7. Assume that one employee of your company spent 2 hours contacting the Concord Company concerning the defective items and arranging to have the goods returned and that this employee is paid $8.50 per hour. An employee earning $9.60 per hour and working in the Accounting Department spent 10 minutes recording the purchase returns. Another employee, who earns $9.20 per hour, spent 15 minutes calculating and recording the adjustment to the invoice as a result of the deduction of the charge for the returned goods. What was the cost to the company in wages for receipt of the damaged goods?

8. Apart from the additional cost to your company in employee wages, what other ramifications of the receipt of the damaged goods from the Concord Company might there be?

9. Apply the series discount and the cash terms offered by the Lexington Company to the invoice received from the Concord Company. Assume that transportation charges are the same, none of the goods are defective, and the bill is paid by November 1. What is the total amount paid? What is the difference between this amount and the amount that would be due the Concord Company if there had been no damaged goods and the bill was paid by November 1? Use your answer to Question 7 to determine whether this difference is greater than the additional cost to the company for employee wages related to the receipt of damaged goods.

10. Would you consider not ordering from the Concord Company in the future? Would it affect your decision if your company generally did not pay an invoice within 10 days but often took advantage of the 15-day cash discount period? What if the transportation charges for shipment from the Lexington Company were less? What if they were greater?

CHAPTER SUMMARY

Key Words

The **list price** of an item is the suggested retail price at which it is listed in a catalogue distributed to dealers. (Objective 7.1A)

A **trade discount** is the amount subtracted from the list price of an item and is expressed as a percent of the list price. (Objective 7.1A)

The **net price** is the amount paid by a dealer for a product. (Objective 7.1A)

A **series discount,** or **chain discount,** consists of two or more trade discounts given on the same item. (Objective 7.1B)

A **single-discount equivalent** is the single discount rate that is equivalent to a series discount. (Objective 7.1C)

Cash discounts are reductions in the net price of a product sold on credit and are offered to encourage early payment of a credit charge. A cash discount is usually a percent of the net price. (Objective 7.2A)

An **invoice** is a bill, or a printed record of the transaction between buyer and seller. (Objective 7.2A)

Among the terms given in an invoice are
    n (net)
    EOM (end of month)
    ROG (receipt of goods)
    2/10, n/30 (two ten, net thirty; a **2**% cash discount is given if the invoice is paid within **10** days; the buyer has **30** days in which to pay the invoice in full.)
    (Objective 7.2A)

Essential Rules

Discount = list price × discount rate (Objective 7.1A)

Net price = list price – discount (Objective 7.1A)

Net price = list price × the complement of the discount rate (Objective 7.1A)

$$\text{List price} = \frac{\text{net price}}{\text{complement of the trade discount rate}}$$ (Objective 7.1A)

Net price after a series discount
  = list price × product of the complements of the discount rates
  (Objective 7.1B)

Single-discount equivalent =
  100% – product of the complements of the discount rates (Objective 7.1C)

Cash discount = net price × percent discount (Objective 7.2A)

Amount due = net price – cash discount (Objective 7.2A)

Amount due = net price × complement of the cash discount rate
(Objective 7.2A)

Partial payment = amount credited × complement of the discount rate
(Objective 7.2B)

Amount due = amount of invoice – amount credited (Objective 7.2B)

To find the due date of an invoice, use the number of days in each month.
    30-day months: April, June, September, November
    31-day months: January, March, May, July, August, October, December
    28-day month: February (Exception: 29 days in a leap year.) (Objective 7.2A)

REVIEW / TEST

1. Find the complement of a 40% trade discount. Express the complement (a) as a percent and (b) as a decimal.

2. A milling machine with a list price of $6200 is advertised at a trade discount of 30%. Find (a) the discount and (b) the net price of the milling machine.

3. A manufacturer's catalogue lists the price of a roof rack as $150. A wholesaler can purchase 100 of the roof racks with a 25% trade discount. Find the net price of the 100 roof racks.

4. One wholesaler lists a tool set at $700 with a discount rate of 35%. Another wholesaler lists a comparable item at $600 with a discount rate of 20%. Which is the better buy?

5. The net price of a printer stand with a 15% discount is $59.50. Find the list price.

6. A video camera with a list price of $950 is advertised at a series discount of 20/10. Find (a) the discount and (b) the net price of the video camera.

7. A 16-foot sailboat is listed at $9400 with a series discount of 30/10. Find the net price of the sailboat.

8. The manager of a shoe store orders $3600 of merchandise and receives a series discount of 25/15. Find the net price of the order.

9. The list price of a card table and four chairs is $170 with a series discount of 20/10. What is the amount of the second trade discount?

10. A manufacturer's list price for a typewriter is $420 with a series discount of 20/15/5. Find the net price.

11. Find the single-discount equivalent of a 35/15/5 series discount. Express the discount (a) as a percent and (b) as a decimal.

12. A clothing store orders $5500 of merchandise and receives a series discount of 25/10. Find the discount.

13. The list price of an electric screwdriver is $25. A series discount of 25/10/5 is offered on all orders greater than $1500. Find the discount for an order of 75 electric screwdrivers.

14. Two different manufacturers are offering comparable items at the same list price, the first with a series discount of 35/15/10 and the second with a series discount of 30/20/10. Which is the better buy?

1. (a) _____

   (b) _____

2. (a) _____

   (b) _____

3. _____

4. _____

5. _____

6. (a) _____

   (b) _____

7. _____

8. _____

9. _____

10. _____

11. (a) _____

    (b) _____

12. _____

13. _____

14. _____

15. An invoice of $4800 is dated March 8, and terms of 2/10, n/30 are offered. Find the discount if the invoice is paid within 10 days.

16. An invoice of $2350 is dated April 13, and terms of 3/10, n/30 are offered. Find the total amount due if the invoice is paid on April 22.

17. A Zenith dealer receives an invoice of $1500 dated May 25, and terms of 4/10, n/30 are offered. Find (a) the cash discount and (b) the amount due if the invoice is paid within 10 days.

18. A Quasar dealer receives an invoice of $16,280 dated August 24, and terms of 2/10, n/30 EOM are offered. Find (a) the last date the discount may be taken, (b) the due date of the invoice, and (c) the amount due if the discount is earned.

19. An invoice of $8240 is dated July 22, and terms of 4/10, n/30 EOM are offered. A payment of $5000 is made on July 30 and the remainder is paid on August 25. Find the total amount paid.

20. An invoice of $4200 is dated June 12, and terms of 3/10, n/30 are offered. A payment of $2500 is made on June 20, and the remainder is paid on July 10. Find the total amount paid.

The following invoice, dated November 12, offers a series discount of 20/10 and a cash discount of 2/10, n/30. The invoice is paid on November 20. Calculate the extensions, the total list price, the series discount, the net price, the cash discount, and the amount to be paid.

15. _____

16. _____

17. (a) _____

   (b) _____

18. (a) _____

   (b) _____

   (c) _____

19. _____

20. _____

21. _____

22. _____

23. _____

24. _____

25. _____

26. _____

27. _____

28. _____

29. _____

30. _____

Quantity	Description	Unit Price	Amount
24	3-piece adjustable wrench set	29.90	21.
18	5-piece plier set	31.80	22.
30	Tool box	13.50	23.
10	10-drawer tool storage set	181.30	24.
40	Belt sander	74.50	25.
		Total List Price	26.
		Trade Discount	27.
		Net Price	28.
		Cash Discount	29.
		Total Amount Paid	30.

8

Pricing

*T*HERE ARE A number of different methods of pricing products. Two such methods are cost-based pricing and competition-based pricing.

A seller who uses **cost-based pricing** first determines the total cost of producing or purchasing a product and then adds an amount to that cost to cover additional expenses and to earn a profit. Markup pricing, which we will examine in this chapter, is a cost-based method of pricing.

Under **competition-based pricing,** a seller uses competitors' prices as a guide in determining the prices to charge for goods and services. This method is most applicable to products that are very similar.

There are also several different pricing strategies that sellers may use. These include odd pricing, multiple-unit pricing, and prestige pricing.

Have you noticed that price tags tend to read $8.99, for example, rather than $9? This is an example of **odd pricing,** which is the strategy of setting a price slightly below a whole number of dollars. Odd pricing is extended to higher-priced items as well. For example, the price of a car may be advertised as $19,995 rather than $20,000. This strategy is based on the belief that consumers respond more positively to the product because they see it as having a lower price.

When shopping at the supermarket, you have probably noticed items marked "2 for 99¢" or "3 for $1.00." These are examples of **multiple-unit pricing,** the practice of setting a single price for two or more items. This strategy increases sales when customers respond to the single price and purchase the multiple units.

High-quality items are usually more expensive than similar items of lesser quality. Therefore, many consumers have come to associate high price with good quality. This is especially true of products such as jewelry, perfume, and cosmetics. **Prestige pricing** reflects the belief that setting a high price on a product imbues it with status and high quality.

Which method of pricing do you think the airline industry employs?[1] Which pricing strategy do you think is applied to a Ford Escort?[2] to a Porsche?[3]

[1]Competition-based pricing

[2]Odd pricing

[3]Prestige pricing

SECTION 8.1   **Markup and Markdown**

***Objective* 8.1A**   *To solve markup problems*

Cost is the amount a merchandising business or retailer pays for a product. **Selling price,** or **retail price,** is the price for which a merchandising business or retailer sells a product to a customer. The difference between selling price and cost is called **markup.** Markup is added to cost to cover the expenses of operating a business and provide a profit to the owners.

Markup can be expressed as a percent of the cost, or it can be expressed as a percent of the selling price. The percent markup is called the **markup rate,** and it must always be expressed as the **markup based on the cost** or the **markup based on the selling price.**

The markup equations when markup is based on cost are

$$M = S - C$$ M = markup
$$M = r \times C$$ S = selling price
$$S = (1 + r)C$$ C = cost
                         r = markup rate

The manager of a clothing store buys a suit for $80 and sells the suit for $116. Find the markup rate on the cost.

◆ Find the markup by solving the $M = S - C$
formula $M = S - C$ for M. $= 116 - 80$
$S = 116$, $C = 80$ $= 36$

◆ Find the markup rate by solving $M = r \times C$
the formula $M = r \times C$ for r. $36 = r \times 80$
$M = 36$, $C = 80$ $\dfrac{36}{80} = r$
                                             $0.45 = r$

The markup rate on the cost is 45%.

A retailer may know from past experience how much customers are willing to pay for a product. Knowing the suggested selling price and the markup rate, the retailer can calculate the maximum cost to be paid for the product.

The manager of a furniture store uses a markup rate of 45% of the cost. If the selling price of a chair is $174, what must be the cost of the chair?

◆ To find the cost, solve the formula $S = (1 + r)C$
$S = (1 + r)C$ for C. $S = 174$, $r = 45\%$ $174 = (1 + 0.45)C$
                                              $174 = 1.45C$
                                              $\dfrac{174}{1.45} = C$
                                              $120 = C$

The cost of the chair is $120.

The markup equations when markup is based on the selling price are:

$$M = S - C$$
$$M = r \times S$$
$$C = (1 - r)S$$

M = markup
S = selling price
C = cost
r = markup rate

A retailer buys a shipment of shirts for \$11.40 each. The retailer uses a markup rate of 40%, based on the selling price. Find the selling price of each shirt.

◆ To find the selling price, solve the formula $C = (1 - r)S$ for S. $C = 11.40$, $r = 40\%$

$$C = (1 - r)S$$
$$11.40 = (1 - 0.40)S$$
$$11.40 = 0.60S$$
$$\frac{11.40}{0.60} = S$$
$$19 = S$$

The selling price is \$19.

Boat Shoes
\$72.00

A shoe store is selling a pair of shoes for \$72 with a markup rate of 30% based on the selling price. Find the cost of the shoes.

◆ To find the cost, solve the formula $C = (1 - r)S$ for C. $S = 72$, $r = 30\%$

$$C = (1 - r)S$$
$$= (1 - 0.30)72$$
$$= (0.70)72$$
$$= 50.40$$

The cost of the shoes is \$50.40.

The difference in the markup rate when it is based on cost and when it is based on selling price can be illustrated by the following: A portable generator cost a hardware store \$112. The selling price of the generator is \$140. Find the markup rate on the cost and the markup rate on the selling price.

◆ To find the markup, solve the formula $M = S - C$ for M. $S = 140$, $C = 112$

$$M = S - C$$
$$= 140 - 112$$
$$= 28$$

◆ To find the markup rate on the cost, solve the formula $M = r \times C$ for r. $M = 28$, $C = 112$

$$M = r \times C$$
$$28 = r \times 112$$
$$\frac{28}{112} = r$$
$$0.25 = r$$

The markup rate on the cost is 25%.

◆ To find the markup rate on the selling price, solve the formula $M = r \times S$ for r. $M = 28$, $S = 140$

$$M = r \times S$$
$$28 = r \times 140$$
$$\frac{28}{140} = r$$
$$0.20 = r$$

The markup rate on the selling price is 20%.

Note that the markup rate on the cost is greater than the markup rate on the selling price. This is always true because the cost is always less than the selling price.

Example 1

A software package costing $219 is sold for $349. Find the markup.

Strategy

To find the markup, solve the formula $M = S - C$ for M. $S = 349$, $C = 219$

Solution

$M = S - C$
$= 349 - 219$
$= 130$

The markup is $130.

Example 2

The cost of an electronic keyboard is $140 and the markup is $56. Find the selling price.

Strategy

To find the selling price, solve the formula $M = S - C$ for S. $M = 56$, $C = 140$

Solution

$M = S - C$
$56 = S - 140$
$196 = S$

The selling price is $196.

Example 3

A pair of crosstrainers with a selling price of $120 has a markup of 30% of the selling price. Find the markup.

Strategy

To find the markup, solve the formula $M = r \times S$ for M. $r = 30\%$, $S = 120$

Solution

$M = 0.30 \times 120$
$= 36$

The markup is $36.

You Try It 1

A treadmill that cost $499 is selling for $698. Find the markup.

Your strategy

Your solution

You Try It 2

The cost of a set of patio furniture is $395. The markup is $154. Find the selling price.

Your strategy

Your solution

You Try It 3

A leather jacket with a selling price of $260 has a markup of 40% of the selling price. Find the markup.

Your strategy

Your solution

Solutions on p. A26

Example 4

A Kodak camera costing $175 is sold for $319. Find the markup rate based on the cost. Round to the nearest tenth of a percent.

Strategy

To find the markup rate:

* Solve the formula $M = S - C$ for M.
  $S = 319$, $C = 175$

* Solve the formula $M = r \times C$ for r.

Solution

$$M = S - C \qquad\qquad M = r \times C$$
$$ = 319 - 175 \qquad 144 = r \times 175$$
$$ = 144 \qquad\qquad \frac{144}{175} = r$$
$$ 0.823 \approx r$$

The markup rate on cost is 82.3%.

You Try It 4

A gas furnace costing $320 is sold for $530. Find the markup rate based on the cost. Round to the nearest tenth of a percent.

Your strategy

Your solution

Example 5

A refrigerator costing $269 is sold for $489. Find the markup rate based on the selling price. Round to the nearest tenth of a percent.

Strategy

To find the markup rate:

* Solve the formula $M = S - C$ for M.
  $C = 269$, $S = 489$

* Solve the formula $M = r \times S$ for r.

Solution

$$M = S - C \qquad\qquad M = r \times S$$
$$ = 489 - 269 \qquad 220 = r \times 489$$
$$ = 220 \qquad\qquad \frac{220}{489} = r$$
$$ 0.450 \approx r$$

The markup rate on selling price is 45.0%.

You Try It 5

An Evinrude outboard motor costing $550 is sold for $900. Find the markup rate based on the selling price. Round to the nearest tenth of a percent.

Your strategy

Your solution

Solutions on pp. A26–A27

Objective 8.1B *To solve markdown problems*

A retailer may reduce the regular price of a product for a promotional sale or because the goods are damaged, odd sizes or colors, or discontinued items. **Markdown** is the amount by which a retailer reduces the regular price of a product. The percent markdown is called the **markdown rate** and is usually expressed as a percent of the original selling price.

The markdown equations are

$$M = R - S$$
$$M = r \times R$$
$$S = (1 - r)R$$

M = markdown
S = selling price
R = regular price
r = markdown rate

In a garden supply store, the regular selling price of a 100-foot garden hose is $32. During an "after-summer sale," the hose is on sale for 25% off the regular price. Find the sale price.

◆ Solve the formula $S = (1 - r)R$ for S.
$R = 32, r = 25\%$

$$S = (1 - r)R$$
$$= (1 - 0.25)32$$
$$= (0.75)32$$
$$= 24$$

The sale price is $24.

Example 6

A lawn mower with a regular price of $579 is on sale for $489. Find the markdown rate. Round to the nearest tenth of a percent.

Strategy

To find the markdown rate:

◆ Solve the formula $M = R - S$ for M.
$R = 579, S = 489$

◆ Solve the formula $M = r \times R$ for r.

Solution

$M = R - S$ $M = r \times R$
$= 579 - 489$ $90 = r \times 579$
$= 90$ $\dfrac{90}{579} = r$
                 $0.155 \approx r$

The markdown rate is 15.5%.

You Try It 6

A garage door opener with a regular price of $159 is on sale for $125. Find the markdown rate. Round to the nearest tenth of a percent.

Your strategy

Your solution

Solution on p. A27

Example 7

A Craftsman belt sander with a regular price of $48 is on sale for 24% off the regular price. Find the sale price.

Strategy

To find the sale price, solve the formula $S = (1 - r)R$ for S. $r = 24\%$, $R = 48$

Solution

$S = (1 - r)R$
$= (1 - 0.24)48$
$= (0.76)48$
$= 36.48$

The sale price is $36.48.

You Try It 7

A pair of Bushnell binoculars with a regular price of $120 is on sale for 35% off the regular price. Find the sale price.

Your strategy

Your solution

Example 8

A gallon of Sherwin-Williams latex semigloss enamel paint is on sale for $12.96 after a markdown of 28%. Find the regular price.

Strategy

To find the regular price, solve the formula $S = (1 - r)R$ for R. $S = 12.96$, $r = 28\%$

Solution

$S = (1 - r)R$
$12.96 = (1 - 0.28)R$
$12.96 = 0.72R$
$\dfrac{12.96}{0.72} = R$
$18 = R$

The regular price is $18.

You Try It 8

A leather attaché is on sale for $148.50 after a markdown of 25%. Find the regular price.

Your strategy

Your solution

Solutions on p. A27

EXERCISE 8.1A

Solve.

1. A barbecue grill costing $244 is sold for $399. Find the markup.

2. A sweater costing $12.50 is sold for $24.95. Find the markup.

3. The cost of a bench saw is $88 and the markup is $57. Find the selling price.

4. The cost of a pair of Sorel boots is $88.25 and the markup is $44.50. Find the selling price.

5. The selling price of a space heater is $239 and the markup is $85. Find the cost.

6. The selling price of a suit is $229.90 and the markup is $96. Find the cost.

7. A bicycle costing $94 has a markup rate of 30% of the cost. Find the markup.

8. An exerciser costing $195 has a markup rate of 42% of the cost. Find the markup.

9. A Black and Decker food processor with a selling price of $149 has a markup of 30% of the selling price. Find the markup.

10. A gun cabinet with a selling price of $429 has a markup of 35% of the selling price. Find the markup.

11. A Whirlpool dishwasher costing $345 has a markup rate of 40% of the cost. Find the selling price.

12. A JVC car stereo costing $155 has a markup rate of 60% of the cost. Find the selling price.

1. _____
2. _____
3. _____
4. _____
5. _____
6. _____
7. _____
8. _____
9. _____
10. _____
11. _____
12. _____

13. A ceiling fan costs a retailer $88. The retailer uses a markup rate of 45% based on cost. Find the selling price of the fan.

14. A retailer uses a markup rate of 25% on a solar light that cost $75. Find the selling price.

15. A Craftsman sander costing $22 has a markup rate of 28% of the selling price. Find the selling price.

16. A home security system costing $172 has a markup rate of 44% of the selling price. Find the selling price.

17. A retailer uses a markup rate of 36% of the selling price. Find the selling price of a mattress that cost the retailer $249.

18. A retailer uses a markup rate of 38% of the selling price. Find the selling price of a clock that cost the retailer $107.

19. An entertainment center selling for $599 has a markup rate of 20% of the selling price. Find the cost.

20. A lawn mower selling for $399 has a markup rate of 30% of the selling price. Find the cost.

21. A retailer uses a markup rate of 24% of the selling price. Find the cost of a set of luggage that the retailer is selling for $179.

22. Find the cost of a sweatshirt that is selling for $39 and has a markup rate of 34% of the selling price.

23. A Hoover vacuum cleaner selling for $189 has a markup rate of 45% of the cost. Find the cost.

24. A guitar selling for $399 has a markup rate of 48% of the cost. Find the cost.

25. Find the cost of an antenna that is selling for $78.88 and has a markup rate of 36% of the cost.

13. _____

14. _____

15. _____

16. _____

17. _____

18. _____

19. _____

20. _____

21. _____

22. _____

23. _____

24. _____

25. _____

26. Find the cost of a video game that is selling for $69.60 and has a markup rate of 45% of the cost.

27. An Amana freezer costing $280 is sold for $394. Find the markup rate on the cost. Round to the nearest tenth of a percent.

28. A water heater with a cost of $95 is sold for $133. Find the markup rate on the cost.

29. A Polaroid camera costing $234 is sold for $390. Find the markup rate on the selling price.

30. A clothes dryer with a cost of $219 is sold for $292. Find the markup rate on the selling price.

26. _____

27. _____

28. _____

29. _____

30. _____

31. _____

32. _____

33. _____

34. _____

35. _____

36. _____

37. _____

*E*XERCISE 8.1B

Solve.

31. A jacket with a regular price of $89 is on sale for $69. Find the markdown.

32. A Magnavox television with a regular price of $599 is on sale for $499. Find the markdown.

33. A JVC stereo with a regular price of $695 is on sale for $557. Find the markdown rate. Round to the nearest tenth of a percent.

34. A Canon typewriter with a regular price of $380 is on sale for $323. Find the markdown rate.

35. A gold necklace is on sale for $156. The regular price of the necklace is $390. Find the markdown rate.

36. The regular price of a gold ring is $215. The ring is on sale for $129. Find the markdown rate.

37. A light fixture with a regular price of $95 is on sale for $61.75. Find the markdown rate.

38. A parka with a regular price of $129 is on sale for $96.75. Find the markdown rate.

39. A corner hutch with a regular price of $299 is on sale for 30% off the regular price. Find the sale price.

40. A Bulova mantel clock with a regular price of $200 is on sale for 18% off the regular price. Find the sale price.

41. The regular price of a dining set is $795. The set is on sale for 20% off the regular price. Find the sale price.

42. Find the sale price of an armchair that has a regular price of $232 and is on sale for 15% off the regular price.

43. A Burlington dining table with a regular price of $375 is on sale for 25% off the regular price. Find the sale price.

44. A battery with a regular price of $75 is on sale for 18% off the regular price. Find the sale price.

45. A tool set is on sale for $124 after a markdown of 20% off the regular price. Find the regular price.

46. A telescope is on sale for $180 after a markdown of 40% off the regular price. Find the regular price.

47. A pair of Dexter shoes is on sale for $78 after a markdown of 22% off the regular price. Find the regular price.

48. A sports jacket is on sale for $49 after a markdown of 20% off the regular price. Find the regular price.

49. A Texas Instruments calculator, marked down 60%, is on sale for $45. Find the regular price.

50. A Seiko wrist watch, marked down 40%, is on sale for $120. Find the regular price.

38. _____

39. _____

40. _____

41. _____

42. _____

43. _____

44. _____

45. _____

46. _____

47. _____

48. _____

49. _____

50. _____

Perishables

Objective 8.2A   *To price perishable items*

Perishables are items that are likely to spoil easily; for example, fruits, vegetables, bakery products, and flowers are perishables. Retailers that sell perishable items generally know from past experience approximately how much of their merchandise will have to be discarded due to spoilage.

Retailers who have paid for any perishables that spoil do not receive any revenue from these goods. If retailers used the same markup for perishables as for nonperishables, they would not earn enough to cover expenses and make a profit. Therefore, a different method of determining markup is used for perishables. The method by which perishables are priced is illustrated in the example below.

Suppose a retailer purchases 500 pounds of oranges for $.20 per pound. The retailer wants a markup rate of 50% based on cost and expects 8% of the oranges to become nonsaleable due to spoilage. What price should the retailer charge per pound for the oranges in order to receive the desired profit?

- ◆ Find the retailer's total cost for the 500 pounds of oranges.

$$\text{Total cost} = \text{unit cost} \times \text{number of units}$$
$$= 0.20 \times 500$$
$$= 100$$

The total cost of the oranges is $100.

- ◆ Find the total selling price if all the oranges are sold.
  $r = 50\% = 0.50, C = 100$

$$S = (1 + r)C$$
$$= (1 + 0.50)100 = (1.50)100 = 150$$

The total selling price without spoilage is $150.

- ◆ Find the percent of the oranges that is expected to be sold.

$$100\% - 8\% = 92\%$$

92% of the oranges are expected to be sold.

- ◆ Use the basic percent equation to find the number of pounds of oranges that is expected to be sold.
  $B = 500, r = 92\% = 0.92$

$$P = B \times R$$
$$= 500 \times 0.92 = 460$$

460 pounds of oranges are expected to be sold.

- ◆ Divide the total selling price without spoilage by the number of pounds expected to be sold. Round to the nearest cent.

$$\frac{\text{total selling price without spoilage}}{\text{number of pounds expected to be sold}} =$$

$$\frac{150}{460} \approx 0.33$$

The retailer should charge $.33 per pound for the oranges.

Example 1

The owner of a bakery wants a 60% markup on cost and expects 5% of the goods prepared to become stale. Doughnuts cost $2.20 per dozen, and 30 dozen doughnuts are baked each day. What price should the owner charge for one dozen doughnuts?

Strategy

To find the price per dozen:

* Find the total cost for 30 dozen doughnuts.

* Find the total selling price if all the doughnuts are sold.

* Find the percent of the doughnuts that is expected to be sold.

* Use the basic percent equation to find the number of dozen doughnuts that is expected to be sold.

* Divide the total selling price without spoilage by the number of dozen expected to be sold. Round to the nearest cent.

Solution

Total cost = unit cost × number of units
$$= 2.20 \times 30$$
$$= 66$$

$$S = (1 + r)C$$
$$= (1 + 0.60)66$$
$$= (1.60)66$$
$$= 105.60$$

$$100\% - 5\% = 95\%$$

$$P = B \times R$$
$$= 30 \times 0.95 = 28.5$$

$$\frac{\text{total selling price without spoilage}}{\text{number of dozen expected to be sold}} = \frac{105.60}{28.5}$$
$$\approx 3.71$$

The price per dozen doughnuts should be $3.71.

You Try It 1

Tomatoes cost a retailer $.24 per pound. The retailer buys 200 pounds and expects 10% of the tomatoes to spoil. If the retailer wants a 55% markup on cost, what should be the selling price per pound of tomatoes?

Your strategy

Your solution

Solution on p. A28

EXERCISE 8.2A

Solve.

1. A florist purchases 10 dozen roses for $1.20 per dozen. The florist wants a 45% markup based on cost. Find (a) the florist's total cost for the roses and (b) the total selling price if all the roses are sold.

2. A grocer buys 300 pounds of watermelons at a cost of $.15 per pound. The grocer wants a 55% markup based on cost. Find (a) the grocer's total cost for the watermelons and (b) the total selling price if all the watermelons are sold.

3. The cost to a baker for corn muffins is $.23 per muffin. The baker wants a markup rate of 60% based on cost and expects 5% of the muffins not to sell before they are stale. The baker prepares 5 dozen corn muffins. Find (a) the total cost for the muffins, (b) the total selling price if all 5 dozen muffins are sold, and (c) the number of muffins that is expected to be sold.

4. Carnations cost $.40 each. A florist purchases 25 carnations and wants a markup rate of 50% based on cost. The florist expects 4% of the carnations to remain unsold. Find (a) the total cost of the carnations, (b) the total selling price if all 25 carnations are sold, and (c) the number of carnations that is expected to be sold.

5. A retailer purchases 150 quarts of blueberries for $.80 per quart. The retailer expects to sell 94% of the blueberries and wants a markup rate of 65% based on cost. Find (a) the total cost of the blueberries, (b) the total selling price if all the blueberries are sold, (c) the number of quarts of blueberries that are expected to be sold, and (d) the price the retailer should charge per quart for the blueberries.

6. A baker prepares 20 loaves of sourdough bread at a cost of $.65 per loaf. The baker expects to sell 90% of the loaves and wants a markup rate of 60% based on cost. Find (a) the total cost of the bread, (b) the total selling price if all the loaves are sold, (c) the number of loaves of bread that are expected to be sold, and (d) the price the baker should charge for each loaf.

1. (a) _____
   (b) _____
2. (a) _____
   (b) _____
3. (a) _____
   (b) _____
   (c) _____
4. (a) _____
   (b) _____
   (c) _____
5. (a) _____
   (b) _____
   (c) _____
   (d) _____
6. (a) _____
   (b) _____
   (c) _____
   (d) _____

7. The cost to a baker for 100 bagels is $.15 per bagel. The baker expects 8% of the bagels to become stale before they are sold and wants a markup rate of 55% based on cost. Find (a) the total cost of the bagels, (b) the total selling price if all the bagels are sold, (c) the number of bagels that are expected to be sold, and (d) the price the baker should charge per bagel.

7. (a) _____

(b) _____

(c) _____

(d) _____

8. A retailer purchases 200 cantaloupes for $.65 each. The retailer expects 10% of the cantaloupes to become nonsaleable due to spoilage and wants a markup rate of 45% based on cost. Find (a) the total cost of the cantaloupes, (b) the total selling price if all the cantaloupes are sold, (c) the number of cantaloupes that are expected to be sold, and (d) the price the retailer should charge for each cantaloupe.

8. (a) _____

(b) _____

(c) _____

(d) _____

9. _____

9. The cost to a grocer for 300 pounds of carrots is $.38 per pound. The grocer wants a markup rate of 50% based on cost and expects 7% of the carrots to become nonsaleable. What price should the grocer charge per pound for the carrots in order to receive the desired profit?

10. _____

11. _____

12. _____

10. A retailer purchases 400 heads of lettuce at a cost of $.52 per head. The retailer wants a markup rate of 65% based on cost and expects 6% of the lettuce to become nonsaleable. What price should the retailer charge per head of lettuce in order to receive the desired profit?

13. _____

14. _____

11. A florist buys 50 lilies at a cost of $1.10 per lily. The florist wants a markup rate of 55% based on cost and expects 10% of the lilies to remain unsold. What price should the florist charge per lily?

12. The cost to a baker for 150 bulky rolls is $.09 per roll. The baker expects 4% of the bulky rolls to become stale before they are sold and wants a markup rate of 60% based on cost. Find the price the baker should charge per roll.

13. The owner of a coffee shop purchases 8 dozen bran muffins at a cost of $3.80 per dozen. The owner wants a markup rate of 70% based on cost and expects that one-half dozen of the muffins will remain unsold. Find the selling price per muffin.

14. The cost to a florist for 5 dozen sweetheart roses is $6 per dozen. The florist wants a markup rate of 50% based on cost and expects that one-half dozen of the roses will remain unsold. Find the selling price per sweetheart rose.

CALCULATORS

% Key The % key on a business calculator can be used for calculations involving markups and markdowns.

In Example 7 on page 220, a belt sander with a regular price of $48 is on sale for 24% off the regular price. The markdown and the sale price can be calculated as follows:

Enter				Display	Comments
48	–	24	%	11.52	The markdown is $11.52.
	=			36.48	The sale price is $36.48.

On some calculators, the sale price is displayed immediately after the % key is pressed. Perform the steps in the example above on your calculator to determine how it operates.

A car stereo costing $185 has a markup rate of 40% of the cost. The markup and the selling price can be calculated as follows:

Enter				Display	Comments
185	+	40	%	74	The markup is $74.
	=			259	The selling price is $259.

A surge protector is selling for $82 with a markup rate of 30% based on the selling price. The markup and the cost of the surge protector can be calculated as follows:

Enter				Display	Comments
82	–	30	%	24.6	The markup is $24.60.
	=			57.4	The cost is $57.40.

When you are using your calculator to solve problems involving amounts of money, set the decimal place selector at "2." Answers will then be rounded to the nearest hundredth, which, when calculating an amount of money, is the nearest cent. The problem below was solved on page 225. Its solution using a calculator follows. In the solution, the calculator was set for 2 decimal places.

A retailer purchases 500 pounds of oranges for $.20 per pound. The retailer wants a markup rate of 50% based on cost and expects 8% of the oranges to become non-saleable due to spoilage. What price should the retailer charge per pound for the oranges in order to receive the desired profit?

Enter						Display	Comments
.2	×	500	+			100.00	The total cost is $100.
50	%	÷				150.00	The total selling price without spoilage is $150.
500	÷	92	%	=		0.33	The selling price is $.33 per pound.

Solve using a business calculator. Round to the nearest cent.

1. A stationery store discounts reams of paper 15% to qualified business customers. Find (a) the markdown and (b) the sale price for a ream of paper that has a regular price of $5.98.

2. A document shredder with a regular price of $149.99 is on sale for 30% off the regular price. Find (a) the markdown and (b) the sale price.

3. A steering wheel bar lock has a markup rate of 55% of the cost. Find (a) the markup and (b) the selling price if the cost of the steering wheel bar lock is $49.98.

4. A Sony pocket color TV costing $99.95 has a markup rate of 45% of the cost. Find (a) the markup and (b) the selling price.

5. A radio-controlled model airplane, regularly priced at $279.75, is on sale for 40% off the regular price. Find (a) the markdown and (b) the sale price.

6. A Panasonic bread machine costs a retailer $112.90. The retailer uses a markup rate of 48% based on cost. Find (a) the markup and (b) the selling price of the bread machine.

7. A paper shredder is selling for $234 with a markup rate of 40% based on the selling price. Find (a) the markup and (b) the cost of the paper shredder.

8. A pair of Reebok Athletic Shoes is priced at $64 with a markup rate of 25% based on the selling price. Find (a) the markup and (b) the cost of the shoes.

9. The owner of a coffee shop purchases 100 muffins at a cost of $.24 per muffin. The owner wants a markup rate of 65% based on cost and expects that 6% of the muffins will remain unsold. Find the selling price per muffin.

10. A retailer purchases 200 pounds of apples for $.35 per pound. The retailer wants a markup rate of 40% based on cost and expects 7% of the apples to become nonsaleable due to spoilage. What price should the retailer charge per pound for the apples in order to receive the desired profit?

1. (a) _____

(b) _____

2. (a) _____

(b) _____

3. (a) _____

(b) _____

4. (a) _____

(b) _____

5. (a) _____

(b) _____

6. (a) _____

(b) _____

7. (a) _____

(b) _____

8. (a) _____

(b) _____

9. _____

10. _____

BUSINESS CASE STUDY

Mail-Order Business

You operate a mail-order book company. One year ago, you decided to initiate a new policy allowing a 10-day free trial on any book purchased from the company. At the end of the 10-day trial period, the customer may either pay cash for the purchase or return the book to the company.

Book sales exhibit seasonal fluctuations; for example, sales are always much higher during November and December, the holiday-buying period. Therefore, in order to accurately evaluate the new policy, you decide to compare the past year's records with the records of the full year of operation before the policy went into effect.

For the year prior to initiation of the new policy, the records show sales of $350,000. Markup was based on selling price, and a markup rate of 35% was used.

During the past year, the first year in which the 10-day free trial was offered to customers, sales totaled $580,000. A markup rate of 35% was still used, and markup continued to be based on selling price.

Because of the increased volume in sales, it was necessary for you to hire two part-time employees. One part-time employee works in the Mail Department packing orders and preparing them for mailing and also assists in restocking items returned by customers. This employee works 20 hours per week and is paid $6.50 per hour. The employee worked 50 weeks during the year.

The second part-time employee assists in the Bookkeeping Department by recording returns in the company ledger. This employee works 10 hours a week and is paid $7.00 per hour. This employee also worked 50 weeks during the year.

In addition to the part-time employees' wages, the company paid a total of $2400 in taxes, insurance, and benefits for the two employees.

Shipping and handling charges on books ordered are paid by the customer if the books are purchased. If the books are returned, these charges are paid by the company. During the first year in which the new policy was in effect, the company paid $18,000 in shipping charges for returned books. The company also paid an additional $14,000 for packaging materials. Your records show that the cost of the books that were neither paid for nor returned was $5000.

1. On the basis of the information provided, is the policy effective? That is, is the company earning greater revenue as a result of allowing a 10-day free trial? Support your position with figures. (*Hint:* Find the difference in markup for the last two years of operation. Subtract from this difference all expenses incurred as a result of the new 10-day free-trial period.)

CHAPTER SUMMARY

Key Words

Cost is the amount a merchandising business or retailer pays for a product. (Objective 8.1A)

Selling price, or **retail price,** is the price for which a merchandising business or retailer sells a product to a customer. (Objective 8.1A)

Markup is the difference between selling price and cost. Markup is added to cost to cover the expenses of operating a business and to provide a profit to the owners. (Objective 8.1A)

Markup can be expressed as a percent of the cost or as a percent of the selling price. The percent markup is called the **markup rate,** and it must always be expressed as the **markup based on the cost** or the **markup based on the selling price.** (Objective 8.1A)

Markdown is the amount by which a retailer reduces the regular price of a product. The percent markdown is called the **markdown rate** and is usually expressed as a percent of the original selling price. (Objective 8.1B)

Perishables are items that are likely to spoil easily; for example, fruits, vegetables, bakery goods, and flowers are perishables. (Objective 8.2A)

Essential Rules

The markup equations when markup is based on cost (Objective 8.1A)

$$M = S - C \qquad\qquad M = \text{markup}$$
$$M = r \times C \qquad\qquad S = \text{selling price}$$
$$S = (1 + r)C \qquad\qquad C = \text{cost}$$
$$\qquad\qquad\qquad\qquad r = \text{markup rate}$$

The markup equations when markup is based on selling price (Objective 8.1A)

$$M = S - C \qquad\qquad M = \text{markup}$$
$$M = r \times S \qquad\qquad S = \text{selling price}$$
$$C = (1 - r)S \qquad\qquad C = \text{cost}$$
$$\qquad\qquad\qquad\qquad r = \text{markup rate}$$

The markdown equations (Objective 8.1B)

$$M = R - S \qquad\qquad M = \text{markdown}$$
$$M = r \times R \qquad\qquad S = \text{selling price}$$
$$S = (1 - r)R \qquad\qquad R = \text{regular price}$$
$$\qquad\qquad\qquad\qquad r = \text{markdown rate}$$

The method of pricing perishable items (Objective 8.2A)

1. Find the retailer's total cost.
   Use the equation Total cost = unit cost × number of units.
2. Find the total selling price if all the items are sold.
   Use the equation $S = (1 + r)C$.
3. Find the amount that is expected to be sold.
4. Divide the total selling price without spoilage by the amount that is expected to be sold.

REVIEW / TEST

1. An air purifier costing $134 is sold for $229. Find the markup.

2. The cost of a chandelier is $159 and the markup is $63.60. Find the selling price.

3. The selling price of a coat rack is $85 and the markup is $39. Find the cost.

4. A lantern costing $59 has a markup rate of 30% of the cost. Find the markup.

5. A microwave oven with a selling price of $180 has a markup rate of 35% of the selling price. Find the markup.

6. An atlas costing $55 has a markup rate of 30% of the cost. Find the selling price.

7. A secretary desk costing $291.20 has a markup rate of 48% of the selling price. Find the selling price.

8. A drum set selling for $699 has a markup rate of 30% of the selling price. Find the cost.

9. Bunk beds selling for $319 have a markup rate of 45% of the cost. Find the cost.

1. _____

2. _____

3. _____

4. _____

5. _____

6. _____

7. _____

8. _____

9. _____

10. A hammock costing $52 is sold for $75.92. Find the markup rate on the cost.

11. A sleeping bag costing $52.93 is sold for $79. Find the markup rate on the selling price.

12. A fishing boat with a regular price of $169 is on sale for $129.99. Find the markdown.

13. An archery set with a regular price of $115 is on sale for $92. Find the markdown rate.

14. A stationery store discounts reams of paper 15% to qualified business customers. Find the sale price of a ream of paper that has a regular price of $5.80.

15. A swing set is on sale for $149.25 after a markdown of 25% off the regular price. Find the regular price.

16. A retailer buys 50 pounds of bean sprouts costing $.56 per pound. The retailer expects that 5 pounds of the bean sprouts will spoil before being sold. A markup rate of 50% based on cost is desired. What price should the retailer charge per pound for the bean sprouts?

17. The cost to a baker for 100 pounds of cookies is $2.15 per pound. The baker expects to sell 92% of the cookies and wants a markup rate of 60% based on cost. What price per pound should the baker charge for the cookies?

18. A grocer purchases 200 pounds of peaches at a cost of $.46 per pound. The grocer expects 10% of the peaches to spoil and wants a markup rate of 55% based on cost. Find the price the grocer should charge per pound of peaches.

10. _____
11. _____
12. _____
13. _____
14. _____
15. _____
16. _____
17. _____
18. _____

9

Payroll

*I*N 1938, THE **Fair Labor Standards Act** was enacted by Congress. This law permits the federal government to set a minimum wage and requires that employees receive overtime pay for working more than 40 hours per week. The first minimum wage was set at $.25 per hour. Today the minimum wage is $4.25 per hour. What percent increase is this?[1]

Not all employees are subject to the Fair Labor Standards Act. For example, managers and professional personnel are exempt from the overtime pay provisions; they are seldom paid overtime for working more than 40 hours per week.

The Fair Labor Standards Act also requires that businesses keep accurate records of each employee. An **employee's individual earnings record** must include wages earned by the employee and the amount of tax withheld from the employee's earnings.

The wages earned by an employee determine not only amounts to be deducted from an employee's paycheck but also amounts the employer must pay to the federal government. For example, an employer is responsible for paying

FICA (Social Security)
FUTA (Federal Unemployment Tax)
SUTA (State Unemployment Tax)

All businesses, no matter how large or small, are required by law to make payments to government agencies by specified deadlines and to submit reports on official forms. The following is a partial list of the reports all employers are responsible for filing.

Employer's Quarterly Federal Tax Return (Form 941)
Employer's Annual Federal Income Tax Report
Transmittal of Income and Tax Statement (Form W-3)
Withholding Statements for Employees (Form W-2)

Because the government imposes penalties if these requirements are not met, and because information required for each of these reports is taken from the employees' individual earnings records, employers take very seriously the responsibility of keeping accurate payroll records.

In some states and cities, employers are also responsible for state and city income taxes. These taxes, too, are based on employee wages and salaries.

In this chapter, you will learn how to calculate gross pay, FICA, federal income tax, and net pay. All of these calculations are vital to the payroll records and earnings reports kept by businesses.

[1]1600%

Gross Pay

Objective 9.1A   *To calculate wages and salaries*

A significant cost of doing business is the cost of employing individuals to carry out the functions of the business. Companies may compensate employees for the work they do by paying them a wage, a salary, a commision, a piecework rate, or a combination of these.

Gross earnings, or **gross pay,** is the total amount an employee has earned. **Net earnings,** or **net pay,** is the "take-home pay" after deductions from the gross earnings have been made.

One of the most common methods of compensation is to pay the employee an hourly wage. Gross pay for an employee paid an hourly wage is found by multiplying the number of hours worked by the hourly rate.

Gross pay = number of hours worked × rate per hour

The majority of companies give overtime pay to employees working more than 40 hours per week. The usual rate for overtime is $1\frac{1}{2}$ or 2 times the regular rate.

Gross pay = regular pay + overtime pay

As shown in the following example, the overtime pay for time and a half is calculated by multiplying the regular hourly wage by 1.5.

A sales clerk worked 48 hours in one week at a regular rate of $5.50 per hour and time and a half for time over 40 hours. Find the sales clerk's gross pay.

◆ Find the regular pay.   Regular pay = 5.50(40) = 220

The regular pay is $220.

◆ Find the overtime pay. The clerk   Overtime pay = 5.50(1.5)(8) = 66
  worked 48 − 40 = 8 hours of
  overtime at time and a half.

The overtime pay is $66.

◆ Add the regular pay to the over-   Gross pay = 220 + 66 = 286
  time pay.

The gross pay is $286.

Information regarding the number of hours worked by employees and their gross earnings is recorded in a book called the **payroll register.** A section of one page from such a register follows.

EMPLOYEE NUMBER	HOURS WORKED						HOURS WORKED		RATE OF PAY	PAY		GROSS EARNINGS
	M	T	W	TH	F	S	REG.	O.T.		REG.	O.T.	
1029	8	8	8	8	8	4	40	4	6.60	264.00	39.60	303.60
1033	$7\frac{1}{2}$	$4\frac{3}{4}$	6	$2\frac{1}{4}$	3		$23\frac{1}{2}$		7.30	171.55		171.55
1117	$9\frac{1}{2}$	8	8	10	8		40	$3\frac{1}{2}$	5.20	208.00	27.30	235.30

Managers and professional employees are usually paid an annual salary, which is paid monthly, semimonthly, weekly, or biweekly.

Monthly—12 paychecks per year
Semimonthly (twice a month)—24 paychecks per year
Weekly—52 paychecks per year
Biweekly (every 2 weeks)—26 paychecks per year

For an employee earning an annual salary, the gross pay per pay period is found by dividing the annual salary by the number of pay periods in a year.

The manager of a fast-food restaurant receives an annual salary of $46,000. Find the gross pay per pay period if the manager is paid monthly.

◆ Divide the annual salary (46,000) by the number of pay periods (12).

$$\text{Gross pay} = 46,000 \div 12 \approx \$3833.33$$

The gross pay per pay period is $3833.33.

Note that gross pay is rounded to the nearest cent.

Example 1

A clerk typist worked 52 hours during one week. The typist receives $5.30 per hour for 40 hours and time and a half for overtime. Determine the typist's gross pay for the week.

You Try It 1

A mason's apprentice worked 48 hours during the week. The apprentice receives $9.30 per hour for 40 hours and time and a half for overtime. Determine the gross pay for the week.

Strategy

To find the gross pay:

◆ Find the regular pay.

◆ Find the overtime pay. The typist worked $52 - 40 = 12$ hours of overtime.

◆ Add the regular pay to the overtime pay.

Your strategy

Solution

Regular pay = 40(5.30) = 212.00

Overtime pay = 5.30(1.5)(12) = 95.40

Gross pay = 212.00 + 95.40 = 307.40

The gross pay is $307.40.

Your solution

Solution on p. A29

Example 2

An accountant earns an annual salary of $38,000. Find the gross pay per pay period if the accountant is paid semimonthly.

Strategy

To find the gross pay per pay period, divide the annual salary (38,000) by the number of pay periods (24).

Solution

Gross pay per period = 38,000 ÷ 24 ≈ 1583.33

The gross pay per pay period is $1583.33.

You Try It 2

A nurse who is paid biweekly earns an annual salary of $32,000. Find the nurse's gross pay per pay period.

Your strategy

Your solution

Solution on p. A29

Objective 9.1B   *To calculate gross pay due to commissions*

Employees involved in sales are often paid a commission. A **commission** is a fixed percent of sales, called the **commission rate,** or a fixed amount per item sold. The employee may be paid on a straight-commission basis or may be paid a combination of salary and commission.

> **Gross pay (commission only) = total sales × commission rate**

> **Gross pay (combination) = salary + commission**

A stock broker receives a weekly salary of $400 and a commission of 2% of sales. Find the gross pay for a week in which the stock broker had sales of $30,000.

◆ Find the commission.   Commission = 30,000 × 0.02 = 600

◆ Add the salary and the   Gross pay = 400 + 600 = 1000
   commission.

The gross pay is $1000.

Some companies provide their salespeople with a **drawing account.** A salesperson can withdraw money from a drawing account. At the end of a pay period, this advance is subtracted from the commission the salesperson earned during the pay period.

To reward top salespersons, a **variable** or **graduated commission scale** may be used. Under this method, different commission rates are applied for different levels of sales.

Use the variable commission rate table at the right to find the gross pay for a salesperson selling $42,500 worth of goods.

Rate Table

Sales	Percent Commission
$0–$25,000	3%
Over $25,000	5%

♦ Add the gross pay for $25,000 in sales to the gross pay for the sales over $25,000 (42,500 − 25,000 = 17,500).

$$25,000 \times 0.03 = 750$$
$$17,500 \times 0.05 = \underline{875}$$
$$\text{Gross pay} = 1625$$

The gross pay is $1625.

A sales representative is allowed a drawing account of $600 per month and receives a monthly commission of 3% of the first $30,000 of sales and 4% of all sales in excess of $30,000. Find the amount due the sales representative for a month in which the representative's drawings were $600 and sales totaled $45,000.

♦ Add the commission for the first $30,000 of sales and the commission for the sales in excess of $30,000 (45,000 − 30,000 = 15,000).

$$30,000 \times 0.03 = 900$$
$$15,000 \times 0.04 = \underline{600}$$
$$\text{Commission} = 1500$$

♦ Subtract the drawings from the commission.

$$1500 - 600 = 900$$

The amount due the sales representative is $900.

Example 3

A shoe clerk receives $4.75 an hour and a commission of 2% of sales. Find the gross pay during a week in which the shoe clerk worked 35 hours and had sales totaling $3250.

Strategy

To find the gross pay:

♦ Multiply the hourly wage (4.75) by the number of hours worked (35).

♦ Multiply the sales (3250) by the commission rate (2% = 0.02).

♦ Add the wages and the commission.

Solution

Hourly wage = 35 × 4.75 = 166.25

Commission = 3250 × 0.02 = 65

Gross pay = 166.25 + 65 = 231.25

The gross pay is $231.25.

You Try It 3

A greeting card manufacturer's representative receives $4.65 an hour and a commission of $3 on each box of cards sold. Find the gross pay for a week in which the representative worked 32 hours and sold 140 boxes of cards.

Your strategy

Your solution

Solution on p. A29

Example 4

A sales representative is allowed a drawing account of $750 per month and receives a monthly commission of 4% of the first $25,000 of sales and 5% of all sales in excess of $25,000. Find the amount due the sales representative for a month in which the representative's drawings were $500 and sales totaled $35,000.

Strategy

To find the amount due:

◆ Add the commission for the first $25,000 of sales and the commission for the sales in excess of $25,000 (35,000 − 25,000 = 10,000) to find the commission.

◆ Subtract the drawings (500) from the commission.

Solution

$25,000 \times 0.04 = 1000$
$10,000 \times 0.05 = \underline{\ \ 500}$
$\text{Commission} = 1500$

$1500 - 500 = 1000$

The amount due the representative is $1000.

You Try It 4

A sales representative is allowed a drawing account of $500 per month and receives a monthly commission of 3% of the first $20,000 of sales and 4% of all sales in excess of $20,000. Find the amount due the sales representative for a month in which the representative's drawings were $400 and sales totaled $55,000.

Your strategy

Your solution

Solution on p. A29

Objective 9.1C   *To calculate gross pay due to piecework*

Employees on the **piecework** plan are paid on the basis of the number of items they produce that pass inspection. This is an incentive basis of pay; the more the worker produces, the higher the wages.

Gross pay = rate per item × number of items produced that pass inspection

Some piecework plans have a quota that the employee must meet, and extra pay is earned for each piece produced beyond the quota.

Some piecework rate structures depend on a tiered scale: as the number of items produced increases, the rate per item paid to the employee also increases. This is referred to as a **differential piece rate.**

A technician receives piecework wages for soldering a circuit board. Payment is arranged as shown at the right. Find the technician's gross earnings for a period in which 200 of the technician's circuit boards pass inspection.

Number of Items Produced	Rate per Item
0–80	$3.50
81–160	$4.50
Over 160	$5.50

◆ Add the pay for completing the first 80 circuit boards, the pay for completing the next 80 circuit boards, and the pay for completing the last 40 circuit boards (200 − 160 = 40).

$$80 \times 3.50 = 280$$
$$80 \times 4.50 = 360$$
$$40 \times 5.50 = \underline{220}$$
$$\text{Gross pay} = 860$$

The gross pay is $860.

Example 5

A lathe operator produces lamp stands and receives $7 for every lamp stand that passes inspection. The operator receives a bonus of $2.50 for all production over 50 lamp stands. Find the gross pay for a period in which 65 lamp stands, two of which were defective, were produced.

Strategy

To find the gross pay:

◆ Find the regular pay by multiplying the regular rate by the number of nondefective lamp stands produced (65 − 2 = 63).

◆ Find the bonus by multiplying the bonus rate by the number produced over 50 that passed inspection (63 − 50 = 13).

◆ Add the regular pay and the bonus.

Solution

Regular pay = $63 \times 7 = 441$

Bonus = $13 \times 2.50 = 32.50$

Gross pay = $441 + 32.50 = 473.50$

The gross pay is $473.50.

You Try It 5

A punch press operator receives 6¢ for every connector pin that passes inspection. The operator receives a bonus of 2¢ for all production over 7500 connector pins. Find the operator's gross pay during a period in which 8400 connector pins were produced and 250 of them were defective.

Your strategy

Your solution

Solution on p. A30

EXERCISE 9.1A

Solve.

1. A secretary worked 35 hours during the week. The secretary earns $7.50 per hour. Determine the secretary's gross pay for the week.

2. A computer operator worked 38 hours during the week. The operator earns $8.75 per hour. Determine the computer operator's gross pay for the week.

3. A welder worked 47 hours during the week. The welder receives $15.50 per hour for 40 hours and time and a half for overtime. Determine the gross pay for the welder.

4. An electrician worked 51 hours during the week. The electrician receives $19.50 per hour for 40 hours and time and a half for overtime. Determine the gross pay for the electrician.

5. A clerk worked 10 hours on Monday, 12 hours on Tuesday, 8 hours on Wednesday, 6 hours on Thursday, 10 hours on Friday, and 5 hours on Saturday. The clerk's rate of pay is $6.50 per hour for 40 hours and time and a half for overtime. Find the clerk's gross pay for the week.

6. A plumber worked 9 hours on Monday, 11 hours on Tuesday, 10 hours on Wednesday, 13 hours on Thursday, 10 hours on Friday, and 4 hours on Saturday. The plumber's rate of pay is $17.25 per hour for 40 hours and time and a half for overtime. Find the plumber's gross pay for the week.

7. A school superintendent receives an annual salary of $62,500. Find the gross pay per period if the superintendent is paid monthly.

8. Jack Brightman receives an annual salary of $78,800. Find Jack's gross monthly pay.

9. A production manager receives an annual salary of $28,400 and is paid biweekly. Find the biweekly gross pay received by the manager.

10. A personnel manager receives an annual salary of $37,750 and is paid biweekly. Find the biweekly gross pay received by the manager.

11. Sally Jacobi receives an annual salary of $42,560. Find Sally's semimonthly pay.

1. _____
2. _____
3. _____
4. _____
5. _____
6. _____
7. _____
8. _____
9. _____
10. _____
11. _____

12. A city manager receives a yearly salary of $59,200 and is paid semimonthly. Find the gross pay received in one pay period.

Complete the following payroll register by finding the number of regular hours worked, the number of overtime hours worked, the regular pay, the overtime pay, and the gross earnings. The overtime rate of pay is one and one half times the regular rate of pay.

HOURS WORKED						HOURS WORKED		RATE OF PAY	PAY		GROSS EARNINGS
M	T	W	TH	F	S	REG.	O.T.		REG.	O.T.	
8	8	8	8	8	8	13.	14.	7.40	15.	16.	17.
4	$6\frac{1}{2}$	7	$5\frac{3}{4}$	4	4	18.	19.	8.20	20.	21.	22.

EXERCISE 9.1B

Solve.

23. A real estate agent receives a commission of 3% of sales. Find the agent's gross pay for selling $124,500 in real estate.

24. An insurance account executive receives a commission of 6% of the first year's premiums. Find the executive's gross pay for selling insurance with $67,200 in first-year premiums.

25. A department store pays a salary of $200 per week plus a 2% commission on sales. Find the weekly pay for (a) a worker who has sales of $15,000 and (b) a worker who has sales of $22,400.

26. A discount store pays a salary of $150 per week plus a 2% commission on sales. Find the weekly pay for (a) a worker who has sales of $11,450 and (b) a worker who has sales of $17,175.

27. A clerk at a clothing store receives a wage of $5.35 per hour and a 5% commission on sales. The clerk sold $2150 of merchandise and worked 36 hours last week. Find the clerk's gross pay for the week.

28. A clerk at a jewelry store receives a wage of $7.25 per hour and a 3% commission on sales. The clerk sold $3250 of merchandise and worked 32 hours last week. Find the clerk's gross pay for the week.

12. _____

13. _____

14. _____

15. _____

16. _____

17. _____

18. _____

19. _____

20. _____

21. _____

22. _____

23. _____

24. _____

25. (a) _____

(b) _____

26. (a) _____

(b) _____

27. _____

28. _____

29. A furniture manufacturer utilizes a variable commission schedule to pay its sales force. The sales staff receives a 3% commission for monthly sales up to $25,000 and 5% for sales over $25,000. Find the gross monthly pay for (a) a worker whose monthly sales are $25,000 and (b) a worker whose monthly sales are $74,000.

30. A wholesaler utilizes a variable commission schedule to pay its sales force. The sales staff receives a 2% commission for monthly sales up to $15,000 and 4% for sales over $15,000. Find the gross monthly pay for (a) a worker whose monthly sales are $35,500 and (b) a worker whose monthly sales are $48,475.

31. A company that sells farm machinery utilizes a variable commission schedule to pay its sales force. The sales staff receives a 2.5% commission on all monthly sales up to $20,000, 4% on the next $30,000, and 6% on all sales over $50,000. Find the gross monthly pay for (a) a worker whose monthly sales are $39,000 and (b) a worker whose monthly sales are $52,000.

32. A company that sells construction materials utilizes a variable commission schedule to pay its sales force. The sales staff receives a 1.5% commission on all monthly sales up to $10,000, 3% on the next $30,000, and 5% on all sales over $40,000. Find the gross monthly pay for (a) a worker whose monthly sales are $42,500 and (b) a worker whose monthly sales are $64,540.

33. Dee Pinckney is paid $4.75 an hour and a commission of $4 on every tool set sold. Find Dee's gross pay for a week in which she worked 25 hours and sold 58 tool sets.

34. Orlando Salavarrio is paid $4.80 an hour and a commission of $3.50 on every can of paint sold. Find Orlando's gross pay for a week in which he worked 36 hours and sold 92 cans of paint.

35. Seiko Hayenga is allowed a drawing account of $800 per month and receives a monthly commission of 2% of the first $15,000 of sales and 4% of all sales in excess of $15,000. Find the amount due Seiko for a month in which her drawings were $800 and her sales totaled $42,000.

36. Norma Holmes is allowed a drawing account of $400 per month and receives a monthly commission of 3% of the first $30,000 of sales and 5% of all sales in excess of $30,000. Find the amount due Norma for a month in which her drawings were $300 and her sales totaled $48,000.

29. (a) _____
(b) _____
30. (a) _____
(b) _____
31. (a) _____
(b) _____
32. (a) _____
(b) _____
33. _____
34. _____
35. _____
36. _____

*E*XERCISE 9.1C

Solve.

37. In an electronics assembly plant, a worker receives $.64 for each circuit board completed. Find the wage for a worker who completes 160 circuit boards.

38. Doug Morgan receives $.52 for each unit produced. Find Doug's gross pay for completing 230 units.

39. A machine operator receives $1.95 per clutch plate. The operator also receives a bonus of $.65 for every clutch plate over 200. Find the gross pay for a period in which 245 clutch plates are produced.

40. A lathe operator receives $4.40 per table base. The operator also receives a bonus of $1.20 for every table base over 150. Find the gross pay for a period in which 230 table bases are produced.

41. An electronics firm pays piecework wages for soldering a circuit board. The weekly pay scale is

1–150 items	$1.50 each
151–200 items	$2.00 each
Over 200 items	$2.50 each

    Find the weekly pay for (a) a worker who completes 180 circuit boards and (b) a worker who completes 262 circuit boards.

42. A farm worker receives piecework wages for picking avocados. The weekly pay scale is

1–80 boxes	$4.00 each
81–120 boxes	$5.20 each
Over 120 boxes	$6.40 each

    Find the weekly pay for (a) a worker who picks 128 boxes of avocados and (b) a worker who picks 143 boxes of avocados.

43. Paul Aquirre receives $6 for every kitchen chair produced that passes inspection. Paul receives a bonus of $1.75 for all production over 45 chairs. Find Paul's gross pay for a period in which he produced 58 kitchen chairs, 4 of which were defective.

44. Amy Strasberg receives $.25 for every plastic car produced that passes inspection. Amy receives a bonus of $.10 for all production over 2500 cars. Find Amy's gross pay for a period in which she produced 3200 plastic cars, 100 of which were defective.

37. _____

38. _____

39. _____

40. _____

41. (a) _____

(b) _____

42. (a) _____

(b) _____

43. _____

44. _____

SECTION 9.2   Deductions

Objective 9.2A   *To calculate FICA*

Deductions are amounts of money that are subtracted from gross pay. Some payroll deductions are required by law—for example, federal income tax and social security tax. Others are requested by the employee for such purposes as health insurance and union dues. A list of representative deductions follows.

Taxes	Retirement Plans
Federal taxes	FICA (social security)
State taxes	State retirement plans
County and city taxes	Company retirement plans
	Private retirement plans

Insurance	Investments	Professional Dues
Life insurance	Savings bonds	Union dues
Disability insurance	Tax-sheltered annuities	Teacher organizations
Medical insurance	Company stock plans	
Dental insurance	Savings in credit unions	

FICA (Federal Insurance Contributions Act), or **social security,** went into effect in 1937 as a supplemental retirement plan. In addition to retirement, it now provides for survivor's benefits, disability insurance, and Medicare. FICA taxes are paid both by the employee and the employer. The employee's share is withheld from the employee's salary and sent to the federal government along with the employer's share.

Each year Congress determines the rate of taxation and the base income subject to FICA tax. For 1993 the FICA rate of taxation was 7.65%. This was separated into two parts: 6.2% for social security and 1.45% for Medicare. The first $57,600 of gross earnings were subject to the social security tax; there was no social security tax on earnings over $57,600. Therefore, the maximum social security tax to be paid by any employee was 6.2% of $57,600, or $3571.20. Only the first $135,000 of gross earnings were subject to the Medicare tax. Therefore, the maximum Medicare tax to be paid by any employee was 1.45% of $135,000, or $1957.50.

The social security administration supplies tax tables to businesses in order to help them determine the proper amount of FICA tax to withhold. A Social Security Tax Table is shown on page 248, and a Medicare Tax Table is shown on page 249. For gross earnings below $57,600, multiplying the employee's earnings by 7.65% yields the same FICA tax to be paid as the figures provided in the Social Security and Medicare tax tables.

TABLE 9.1

Note: *Wages subject to social security are generally also subject to the Medicare tax.*

Wages at least	But less than	Tax to be withheld	Wages at least	But less than	Tax to be withheld	Wages at least	But less than	Tax to be withheld	Wages at least	But less than	Tax to be withheld
$54.12	$54.28	$3.36	$66.54	$66.70	$4.13	$78.96	$79.12	$4.90	$91.38	$91.54	$5.67
54.28	54.44	3.37	66.70	66.86	4.14	79.12	79.28	4.91	91.54	91.70	5.68
54.44	54.60	3.38	66.86	67.02	4.15	79.28	79.44	4.92	91.70	91.86	5.69
54.60	54.76	3.39	67.02	67.18	4.16	79.44	79.60	4.93	91.86	92.02	5.70
54.76	54.92	3.40	67.18	67.34	4.17	79.60	79.76	4.94	92.02	92.18	5.71
54.92	55.09	3.41	67.34	67.50	4.18	79.76	79.92	4.95	92.18	92.34	5.72
55.09	55.25	3.42	67.50	67.67	4.19	79.92	80.09	4.96	92.34	92.50	5.73
55.25	55.41	3.43	67.67	67.83	4.20	80.09	80.25	4.97	92.50	92.67	5.74
55.41	55.57	3.44	67.83	67.99	4.21	80.25	80.41	4.98	92.67	92.83	5.75
55.57	55.73	3.45	67.99	68.15	4.22	80.41	80.57	4.99	92.83	92.99	5.76
55.73	55.89	3.46	68.15	68.31	4.23	80.57	80.73	5.00	92.99	93.15	5.77
55.89	56.05	3.47	68.31	68.47	4.24	80.73	80.89	5.01	93.15	93.31	5.78
56.05	56.21	3.48	68.47	68.63	4.25	80.89	81.05	5.02	93.31	93.47	5.79
56.21	56.38	3.49	68.63	68.80	4.26	81.05	81.21	5.03	93.47	93.63	5.80
56.38	56.54	3.50	68.80	68.96	4.27	81.21	81.38	5.04	93.63	93.80	5.81
56.54	56.70	3.51	68.96	69.12	4.28	81.38	81.54	5.05	93.80	93.96	5.82
56.70	56.86	3.52	69.12	69.28	4.29	81.54	81.70	5.06	93.96	94.12	5.83
56.86	57.02	3.53	69.28	69.44	4.30	81.70	81.86	5.07	94.12	94.28	5.84
57.02	57.18	3.54	69.44	69.60	4.31	81.86	82.02	5.08	94.28	94.44	5.85
57.18	57.34	3.55	69.60	69.76	4.32	82.02	82.18	5.09	94.44	94.60	5.86
57.34	57.50	3.56	69.76	69.92	4.33	82.18	82.34	5.10	94.60	94.76	5.87
57.50	57.67	3.57	69.92	70.09	4.34	82.34	82.50	5.11	94.76	94.92	5.88
57.67	57.83	3.58	70.09	70.25	4.35	82.50	82.67	5.12	94.92	95.09	5.89
57.83	57.99	3.59	70.25	70.41	4.36	82.67	82.83	5.13	95.09	95.25	5.90
57.99	58.15	3.60	70.41	70.57	4.37	82.83	82.99	5.14	95.25	95.41	5.91
58.15	58.31	3.61	70.57	70.73	4.38	82.99	83.15	5.15	95.41	95.57	5.92
58.31	58.47	3.62	70.73	70.89	4.39	83.15	83.31	5.16	95.57	95.73	5.93
58.47	58.63	3.63	70.89	71.05	4.40	83.31	83.47	5.17	95.73	95.89	5.94
58.63	58.80	3.64	71.05	71.21	4.41	83.47	83.63	5.18	95.89	96.05	5.95
58.80	58.96	3.65	71.21	71.38	4.42	83.63	83.80	5.19	96.05	96.21	5.96
58.96	59.12	3.66	71.38	71.54	4.43	83.80	83.96	5.20	96.21	96.38	5.97
59.12	59.28	3.67	71.54	71.70	4.44	83.96	84.12	5.21	96.38	96.54	5.98
59.28	59.44	3.68	71.70	71.86	4.45	84.12	84.28	5.22	96.54	96.70	5.99
59.44	59.60	3.69	71.86	72.02	4.46	84.28	84.44	5.23	96.70	96.86	6.00
59.60	59.76	3.70	72.02	72.18	4.47	84.44	84.60	5.24	96.86	97.02	6.01
59.76	59.92	3.71	72.18	72.34	4.48	84.60	84.76	5.25	97.02	97.18	6.02
59.92	60.09	3.72	72.34	72.50	4.49	84.76	84.92	5.26	97.18	97.34	6.03
60.09	60.25	3.73	72.50	72.67	4.50	84.92	85.09	5.27	97.34	97.50	6.04
60.25	60.41	3.74	72.67	72.83	4.51	85.09	85.25	5.28	97.50	97.67	6.05
60.41	60.57	3.75	72.83	72.99	4.52	85.25	85.41	5.29	97.67	97.83	6.06
60.57	60.73	3.76	72.99	73.15	4.53	85.41	85.57	5.30	97.83	97.99	6.07
60.73	60.89	3.77	73.15	73.31	4.54	85.57	85.73	5.31	97.99	98.15	6.08
60.89	61.05	3.78	73.31	73.47	4.55	85.73	85.89	5.32	98.15	98.31	6.09
61.05	61.21	3.79	73.47	73.63	4.56	85.89	86.05	5.33	98.31	98.47	6.10
61.21	61.38	3.80	73.63	73.80	4.57	86.05	86.21	5.34	98.47	98.63	6.11
61.38	61.54	3.81	73.80	73.96	4.58	86.21	86.38	5.35	98.63	98.80	6.12
61.54	61.70	3.82	73.96	74.12	4.59	86.38	86.54	5.36	98.80	98.96	6.13
61.70	61.86	3.83	74.12	74.28	4.60	86.54	86.70	5.37	98.96	99.12	6.14
61.86	62.02	3.84	74.28	74.44	4.61	86.70	86.86	5.38	99.12	99.28	6.15
62.02	62.18	3.85	74.44	74.60	4.62	86.86	87.02	5.39	99.28	99.44	6.16
62.18	62.34	3.86	74.60	74.76	4.63	87.02	87.18	5.40	99.44	99.60	6.17
62.34	62.50	3.87	74.76	74.92	4.64	87.18	87.34	5.41	99.60	99.76	6.18
62.50	62.67	3.88	74.92	75.09	4.65	87.34	87.50	5.42	99.76	99.92	6.19
62.67	62.83	3.89	75.09	75.25	4.66	87.50	87.67	5.43	99.92	100.00	6.20
62.83	62.99	3.90	75.25	75.41	4.67	87.67	87.83	5.44			
62.99	63.15	3.91	75.41	75.57	4.68	87.83	87.99	5.45			
63.15	63.31	3.92	75.57	75.73	4.69	87.99	88.15	5.46			
63.31	63.47	3.93	75.73	75.89	4.70	88.15	88.31	5.47			
63.47	63.63	3.94	75.89	76.05	4.71	88.31	88.47	5.48			
63.63	63.80	3.95	76.05	76.21	4.72	88.47	88.63	5.49			
63.80	63.96	3.96	76.21	76.38	4.73	88.63	88.80	5.50	**Wages**	**Taxes**	
63.96	64.12	3.97	76.38	76.54	4.74	88.80	88.96	5.51	$100	$6.20	
64.12	64.28	3.98	76.54	76.70	4.75	88.96	89.12	5.52	200	12.40	
64.28	64.44	3.99	76.70	76.86	4.76	89.12	89.28	5.53	300	18.60	
64.44	64.60	4.00	76.86	77.02	4.77	89.28	89.44	5.54	400	24.80	
64.60	64.76	4.01	77.02	77.18	4.78	89.44	89.60	5.55	500	31.00	
64.76	64.92	4.02	77.18	77.34	4.79	89.60	89.76	5.56	600	37.20	
64.92	65.09	4.03	77.34	77.50	4.80	89.76	89.92	5.57	700	43.40	
65.09	65.25	4.04	77.50	77.67	4.81	89.92	90.09	5.58	800	49.60	
65.25	65.41	4.05	77.67	77.83	4.82	90.09	90.25	5.59	900	55.80	
65.41	65.57	4.06	77.83	77.99	4.83	90.25	90.41	5.60	1,000	62.00	
65.57	65.73	4.07	77.99	78.15	4.84	90.41	90.57	5.61			
65.73	65.89	4.08	78.15	78.31	4.85	90.57	90.73	5.62			
65.89	66.05	4.09	78.31	78.47	4.86	90.73	90.89	5.63			
66.05	66.21	4.10	78.47	78.63	4.87	90.89	91.05	5.64			
66.21	66.38	4.11	78.63	78.80	4.88	91.05	91.21	5.65			
66.38	66.54	4.12	78.80	78.96	4.89	91.21	91.38	5.66			

Self-employed individuals must pay a self-employment tax to fund social security and Medicare. Since the contribution by a self-employed person is not matched by an employer, the tax rates are higher. In 1993, the tax rate was 15.3%, with 12.4% applied to social security and 2.9% applied to Medicare. The maximum amount subject to the social security tax was $57,600, and the maximum amount subject to the Medicare tax was $135,000.

TABLE 9.2

1.45% Medicare Employee Tax Table for 1993

Wages at least	But less than	Tax to be withheld	Wages at least	But less than	Tax to be withheld	Wages at least	But less than	Tax to be withheld	Wages at least	But less than	Tax to be withheld
$0.00	$0.35	$0.00	$28.63	$29.32	$.42	$57.59	$58.28	$.84	$86.56	$87.25	$1.26
.35	1.04	.01	29.32	30.00	.43	58.28	58.97	.85	87.25	87.94	1.27
1.04	1.73	.02	30.00	30.69	.44	58.97	59.66	.86	87.94	88.63	1.28
1.73	2.42	.03	30.69	31.38	.45	59.66	60.35	.87	88.63	89.32	1.29
2.42	3.11	.04	31.38	32.07	.46	60.35	61.04	.88	89.32	90.00	1.30
3.11	3.80	.05	32.07	32.76	.47	61.04	61.73	.89	90.00	90.69	1.31
3.80	4.49	.06	32.76	33.45	.48	61.73	62.42	.90	90.69	91.38	1.32
4.49	5.18	.07	33.45	34.14	.49	62.42	63.11	.91	91.38	92.07	1.33
5.18	5.87	.08	34.14	34.83	.50	63.11	63.80	.92	92.07	92.76	1.34
5.87	6.56	.09	34.83	35.52	.51	63.80	64.49	.93	92.76	93.45	1.35
6.56	7.25	.10	35.52	36.21	.52	64.49	65.18	.94	93.45	94.14	1.36
7.25	7.94	.11	36.21	36.90	.53	65.18	65.87	.95	94.14	94.83	1.37
7.94	8.63	.12	36.90	37.59	.54	65.87	66.56	.96	94.83	95.52	1.38
8.63	9.32	.13	37.59	38.28	.55	66.56	67.25	.97	95.52	96.21	1.39
9.32	10.00	.14	38.28	38.97	.56	67.25	67.94	.98	96.21	96.90	1.40
10.00	10.69	.15	38.97	39.66	.57	67.94	68.63	.99	96.90	97.59	1.41
10.69	11.38	.16	39.66	40.35	.58	68.63	69.32	1.00	97.59	98.28	1.42
11.38	12.07	.17	40.35	41.04	.59	69.32	70.00	1.01	98.28	98.97	1.43
12.07	12.76	.18	41.04	41.73	.60	70.00	70.69	1.02	98.97	99.66	1.44
12.76	13.45	.19	41.73	42.42	.61	70.69	71.38	1.03	99.66	100.00	1.45
13.45	14.14	.20	42.42	43.11	.62	71.38	72.07	1.04			
14.14	14.83	.21	43.11	43.80	.63	72.07	72.76	1.05			
14.83	15.52	.22	43.80	44.49	.64	72.76	73.45	1.06			
15.52	16.21	.23	44.49	45.18	.65	73.45	74.14	1.07			
16.21	16.90	.24	45.18	45.87	.66	74.14	74.83	1.08			
16.90	17.59	.25	45.87	46.56	.67	74.83	75.52	1.09			
17.59	18.28	.26	46.56	47.25	.68	75.52	76.21	1.10	Wages	Taxes	
18.28	18.97	.27	47.25	47.94	.69	76.21	76.90	1.11			
18.97	19.66	.28	47.94	48.63	.70	76.90	77.59	1.12	$100	$1.45	
19.66	20.35	.29	48.63	49.32	.71	77.59	78.28	1.13	200	2.90	
20.35	21.04	.30	49.32	50.00	.72	78.28	78.97	1.14	300	4.35	
21.04	21.73	.31	50.00	50.69	.73	78.97	79.66	1.15	400	5.80	
21.73	22.42	.32	50.69	51.38	.74	79.66	80.35	1.16	500	7.25	
22.42	23.11	.33	51.38	52.07	.75	80.35	81.04	1.17	600	8.70	
23.11	23.80	.34	52.07	52.76	.76	81.04	81.73	1.18	700	10.15	
23.80	24.49	.35	52.76	53.45	.77	81.73	82.42	1.19	800	11.60	
24.49	25.18	.36	53.45	54.14	.78	82.42	83.11	1.20	900	13.05	
25.18	25.87	.37	54.14	54.83	.79	83.11	83.80	1.21	1,000	14.50	
25.87	26.56	.38	54.83	55.52	.80	83.80	84.49	1.22			
26.56	27.25	.39	55.52	56.21	.81	84.49	85.18	1.23			
27.25	27.94	.40	56.21	56.90	.82	85.18	85.87	1.24			
27.94	28.63	.41	56.90	57.59	.83	85.87	86.56	1.25			

Find the FICA tax on gross earnings of $264.80 by using the Social Security Tax Table and the Medicare Tax Table.

- Note at the bottom of the Social Security Tax Table on page 248 that the tax on $200 is $12.40. · Social security tax on $200 = 12.40

- Find the social security tax for the remainder ($64.80). Look on page 248; $64.80 is between $64.76 and $64.92. · Social security tax on $64.80 = 4.02

- Note at the bottom of the Medicare Tax Table on this page that the tax on $200 is $2.90. · Medicare tax on $200 = 2.90

- Find the Medicare tax for the remainder ($64.80). $64.80 is between $64.49 and $65.18. · Medicare tax on $64.80 = .94

- Add the four amounts. · Total FICA tax = 20.26

The FICA tax is $20.26.

Find the FICA tax on gross earnings of $264.80 by using a tax rate of 7.65%.

♦ Multiply the gross earnings by   FICA tax = 264.80×0.0765
   7.65%. $= 20.2572$

The FICA tax is $20.26.

Note that this is the same FICA tax that we found by using the tax tables in the previous example.

Whereas some employers list the total FICA tax deducted from the paycheck as one figure, other employers list separately the social security tax deducted and the Medicare tax deducted. For example, the social security tax deducted may be listed under FICA/OASI TAX, where OASI stands for Old Age Survivors Insurance Tax, and the Medicare tax under FICA/MED TAX.

Use the tax rates of 6.2% for social security and 1.45% for Medicare to find the social security tax and the Medicare tax to be deducted from gross earnings of $816.72. The accumulated gross earnings are less than $57,600.

♦ Multiply the gross earnings by   Social
   6.2% to find the social security   security tax = 816.72×0.062
   tax. $= 50.63664$

♦ Multiply the gross earnings by   Medicare tax = 816.72×0.0145
   1.45% to find the Medicare tax. $= 11.84244$

The social security tax is $50.64 and the Medicare tax is $11.84.

An employee earns a gross salary of $4965 per month. Find the social security tax deduction and Medicare tax deduction taken from the employee's December paycheck. Use the tax rates of 6.2% and 1.45%.

♦ Find the gross salary earned   Gross earnings = 11×4965
   during the previous 11 months $= 54,615$
   of the year.

♦ Find the amount of the December   Taxable salary = $57,600 - 54,615$
   check that is subject to social $= 2985$
   security taxes.

♦ Multiply the amount subject to   Social security tax = 2985×0.062
   social security taxes by 6.2%. $= 185.07$

♦ The total earnings for December   Medicare tax = 4965×0.0145
   are subject to Medicare taxes. $= 71.9925$
   Multiply the December earnings
   by 1.45%.

The social security tax deduction is $185.07, and the Medicare tax deduction is $71.99.

Example 1

Find the FICA tax to be deducted from gross earnings of $384.70. The accumulated gross earnings are less than $57,600. Use the tax rate of 7.65%.

Strategy

To find the FICA deduction, multiply the gross earnings (384.70) by 7.65%.

Solution

$384.70 \times 0.0765 = 29.42955$

The FICA tax deduction is $29.43.

You Try It 1

Find the FICA tax to be deducted from gross earnings of $492.30. The accumulated gross earnings are less than $57,600. Use the tax rate of 7.65%.

Your strategy

Your solution

Example 2

Use the Social Security Tax Table and the Medicare Tax Table to find the FICA tax on $2364.50. The accumulated gross earnings are less than $57,600.

Strategy

To find the FICA tax:

- Find the social security tax on $2000 (multiply the tax on $1000 by 2), on $300, and on $64.50.

- Find the Medicare tax on $2000 (multiply the tax on $1000 by 2), on $300, and on $64.50.

- Add the six amounts.

Solution

Social security tax

on $2000 = 2(62.00)	=	124.00
on $300	=	18.60
on $64.50	=	4.00

Medicare tax

on $2000 = 2(14.50)	=	29.00
on $300	=	4.35
on $64.50	=	.94
Total FICA tax	=	180.89

The FICA tax is $180.90.

You Try It 2

Use the Social Security Tax Table and the Medicare Tax Table to find the FICA tax on $3275.40. The accumulated gross earnings are less than $57,600.

Your strategy

Your solution

Solutions on p. A30

Example 3

Shelly Egan earns a gross salary of $5325 per month. Find the total FICA deduction taken from Shelly's November paycheck. Use the tax rates of 6.2% and 1.45%.

Strategy

To find the FICA deduction:

* Find the gross salary earned during the previous 10 months of the year.

* Find the amount of the November check that is subject to social security tax.

* Multiply the amount subject to social security tax by 6.2%.

* Multiply the amount subject to Medicare tax by 1.45%.

* Add the social security tax and the Medicare tax.

Solution

Gross earnings = $10 \times 5325 = 53{,}250$

Taxable salary for social security
$$= 57{,}600 - 53{,}250 = 4350$$

Social security tax $= 4350 \times 0.062$
$$= 269.70$$

Medicare tax $= 5325 \times 0.0145$
$$= 77.2125$$

Total FICA tax $= 269.70 + 77.21$
$$= 346.91$$

The total FICA deduction is $346.91.

You Try It 3

Peter Duval earns a gross salary of $4975 per month. Find the total FICA deduction taken from Peter's December paycheck. Use the tax rates of 6.2% and 1.45%.

Your strategy

Your solution

Solution on p. A31

Objective 9.2B   *To calculate federal income tax*

The FICA tax is the second largest source of revenue for the federal government. The largest source of revenue is the **individual** or **personal income tax.** The personal income tax is a **progressive tax:** as the individual's income increases, an increasing proportion of income is paid as income tax.

Each employee is required to file with the employer a completed W-4 form. This form has the name, address, marital status, and social security number of the employee, as well as the number of exemptions claimed by the employee. The number of exemptions is the number of dependents claimed by the employee and is referred to as the number of **withholding allowances**.

The government allows one withholding allowance for each person supported by the employee. The employee is counted as a dependent. For example, a married person with 2 children might claim 4 dependents: 1 for the employee, 1 for the spouse, and 1 for each child. The instructions on a W-4 form help an employee determine how many allowances to claim.

Form **W-4** Department of the Treasury Internal Revenue Service	**Employee's Withholding Allowance Certificate** ▶ **For Privacy Act and Paperwork Reduction Act Notice, see reverse.**	OMB No. 1545-0010 **1993**

1 Type or print your first name and middle initial   Last name   **2** Your social security number

Home address (number and street or rural route)

3 ☐ Single ☐ Married ☐ Married, but withhold at higher Single rate.
 Note: *If married, but legally separated, or spouse is a nonresident alien, check the Single box.*

City or town, state, and ZIP code

4 If your last name differs from that on your social security card, check here and call 1-800-772-1213 for more information · · · · ▶ ☐

5 Total number of allowances you are claiming (from line G above or from the worksheets on page 2 if they apply) .   **5**

6 Additional amount, if any, you want withheld from each paycheck . . . .   **6** $

7 I claim exemption from withholding for 1993 and I certify that I meet **ALL** of the following conditions for exemption:
- Last year I had a right to a refund of **ALL** Federal income tax withheld because I had **NO** tax liability; **AND**
- This year I expect a refund of **ALL** Federal income tax withheld because I expect to have **NO** tax liability; **AND**
- This year if my income exceeds $600 and includes nonwage income, another person cannot claim me as a dependent.

If you meet all of the above conditions, enter "EXEMPT" here . . . . ▶   **7**

Under penalties of perjury, I certify that I am entitled to the number of withholding allowances claimed on this certificate or entitled to claim exempt status.

Employee's signature ▶   Date ▶   , 19

8 Employer's name and address (Employer: Complete 8 and 10 only if sending to the IRS)   **9** Office code (optional)   **10** Employer identification number

Federal income tax (**FIT**) is withheld from each paycheck. The withholding tax is based on the employee's gross earnings for that pay period. The more allowances claimed on a W-4 form, the less money the employer deducts from a paycheck for FIT. The number of allowances claimed may be increased or decreased, within governmental regulations, by the employee. If in previous years the employee has received a tax refund when filing an income tax return at the end of the year, the number of allowances might be raised so that the employee receives more per pay period. If the employee has owed money at the end-of-the-year filing, the number of exemptions might be lowered so that the employee will pay more federal income tax per pay period.

The federal government provides withholding tables to guide employers in withholding the proper amount of federal income tax. In order to use the withholding tables, it is necessary to know the pay period, the amount of pay, and the number of exemptions. To determine the amount of federal withholding tax to be deducted from a paycheck, one of two methods is used: the percentage method or the wage bracket method. We will discuss the percentage method first and then look at the wage bracket method.

For the **percentage method** of determining the amount of federal income tax to withhold, a Percentage Method Income Tax Withholding Table is used (see Table 9.3). This table provides the withholding allowance to be subtracted from gross pay in order to find the amount subject to withholding. The amount subject to federal withholding is called the **federal earnings.** The Withholding Tax Tables for Percentage Method of Withholding on page 255 (Table 9.4) are used to calculate the amount of withholding tax to be paid.

TABLE 9.3   Percentage Method Income Tax Withholding Table

Payroll Period	One Withholding Allowance
Weekly	$ 45.19
Biweekly	90.38
Semimonthly..	97.92
Monthly	195.83
Quarterly...	587.50
Semiannually..	1,175.00
Annually	2,350.00
Daily or miscellaneous (each day of the payroll period)......................	9.04

Use the percentage method to find the FIT deduction for a married person who claims five exemptions and is paid $5340 per month.

- ◆ Use Table 9.3 to find the amount for one withholding allowance for a monthly payroll period. Multiply the allowance by the number of exemptions claimed.

  Withholding allowance = 5(195.83)
  = 979.15

- ◆ Subtract the withholding allowance from the gross pay to find the federal earnings.

  Federal earnings = 5340 − 979.15
  = 4360.85

- ◆ Use Table 9.4 on page 255 to calculate the withholding tax. These figures are taken from Table 4, Column (b).

  FIT = 431.85 + 0.28(4360.85 − 3396)
  = 431.85 + 0.28(964.85)
  = 431.85 + 270.16
  = 702.01

The FIT deduction is $702.01.

The tables used to determine the federal withholding tax by the **wage bracket method** are available for single and married employees who are paid weekly, biweekly, monthly, semimonthly, and so on. Table 9.5 on page 256 provides deductions for single persons who are paid weekly. Table 9.6 on page 257 provides deductions for married persons who are paid monthly.

Use the wage bracket method to find the FIT deduction for a married person who claims five exemptions and has a monthly income of $5340. (Note that this is the same example we solved above using the percentage method.)

- ◆ Use Table 9.6 on page 257. $5340 is between $5320 and $5360. Look to the right in the column headed 5 withholding allowances.

The FIT deduction is $702.

TABLE 9.4

Tables for Percentage Method of Withholding
(For Wages Paid in 1993)

TABLE 1—WEEKLY Payroll Period

(a) SINGLE person (including head of household)—

If the amount of wages (after subtracting withholding allowances) is:	The amount of income tax to withhold is:
Not over $49	$0

Over—	But not over—		of excess over—
$49	—$451 . . .	15%	—$49
$451	—$942 . . .	$60.30 plus 28%	—$451
$942		$197.78 plus 31%	—$942

(b) MARRIED person—

If the amount of wages (after subtracting withholding allowances) is:	The amount of income tax to withhold is:
Not over $119	$0

Over—	But not over—		of excess over—
$119	—$784 . . .	15%	—$119
$784	—$1,563 . . .	$99.75 plus 28%	—$784
$1,563		$317.87 plus 31%	—$1,563

TABLE 2—BIWEEKLY Payroll Period

(a) SINGLE person (including head of household)—

If the amount of wages (after subtracting withholding allowances) is:	The amount of income tax to withhold is:
Not over $97	$0

Over—	But not over—		of excess over—
$97	—$902 . . .	15%	—$97
$902	—$1,884 . .	$120.75 plus 28%	—$902
$1,884		$395.71 plus 31%	—$1,884

(b) MARRIED person—

If the amount of wages (after subtracting withholding allowances) is:	The amount of income tax to withhold is:
Not over $238	$0

Over—	But not over—		of excess over—
$238	—$1,567 . . .	15%	—$238
$1,567	—$3,125 . . .	$199.35 plus 28%	—$1,567
$3,125		$635.59 plus 31%	—$3,125

TABLE 3—SEMIMONTHLY Payroll Period

(a) SINGLE person (including head of household)—

If the amount of wages (after subtracting withholding allowances) is:	The amount of income tax to withhold is:
Not over $105	$0

Over—	But not over—		of excess over—
$105	—$977 . . .	15%	—$105
$977	—$2,041 . .	$130.80 plus 28%	—$977
$2,041		$428.72 plus 31%	—$2,041

(b) MARRIED person—

If the amount of wages (after subtracting withholding allowances) is:	The amount of income tax to withhold is:
Not over $258	$0

Over—	But not over—		of excess over—
$258	—$1,698 . . .	15%	—$258
$1,698	—$3,385 . . .	$216.00 plus 28%	—$1,698
$3,385		$688.36 plus 31%	—$3,385

TABLE 4—MONTHLY Payroll Period

(a) SINGLE person (including head of household)—

If the amount of wages (after subtracting withholding allowances) is:	The amount of income tax to withhold is:
Not over $210	$0

Over—	But not over—		of excess over—
$210	—$1,954 . .	15%	—$210
$1,954	—$4,081 . .	$261.60 plus 28%	—$1,954
$4,081		$857.16 plus 31%	—$4,081

(b) MARRIED person—

If the amount of wages (after subtracting withholding allowances) is:	The amount of income tax to withhold is:
Not over $517	$0

Over—	But not over—		of excess over—
$517	—$3,396 . . .	15%	—$517
$3,396	—$6,771 . . .	$431.85 plus 28%	—$3,396
$6,771		$1,376.85 plus 31%	—$6,771

TABLE 9.5   Table for Wage Bracket Method of Withholding

		(For Wages Paid in 1993)										

If the wages are—		And the number of withholding allowances claimed is—										
At least	But less than	0	1	2	3	4	5	6	7	8	9	10
		The amount of income tax to be withheld is—										
$590	$600	$101	$88	$75	$63	$55	$48	$41	$35	$28	$21	$14
600	610	103	91	78	66	56	50	43	36	29	22	16
610	620	106	94	81	68	58	51	44	38	31	24	17
620	630	109	96	84	71	59	53	46	39	32	25	19
630	640	112	99	87	74	61	54	47	41	34	27	20
640	650	115	102	89	77	64	56	49	42	35	28	22
650	660	117	105	92	80	67	57	50	44	37	30	23
660	670	120	108	95	82	70	59	52	45	38	31	25
670	680	123	110	98	85	72	60	53	47	40	33	26
680	690	126	113	101	88	75	63	55	48	41	34	28
690	700	129	116	103	91	78	65	56	50	43	36	29
700	710	131	119	106	94	81	68	58	51	44	37	31
710	720	134	122	109	96	84	71	59	53	46	39	32
720	730	137	124	112	99	86	74	61	54	47	40	34
730	740	140	127	115	102	89	77	64	56	49	42	35
740	750	143	130	117	105	92	79	67	57	50	43	37
750	760	145	133	120	108	95	82	70	59	52	45	38
760	770	148	136	123	110	98	85	72	60	53	46	40
770	780	151	138	126	113	100	88	75	63	55	48	41
780	790	154	141	129	116	103	91	78	65	56	49	43
790	800	157	144	131	119	106	93	81	68	58	51	44
800	810	159	147	134	122	109	96	84	71	59	52	46
810	820	162	150	137	124	112	99	86	74	61	54	47
820	830	165	152	140	127	114	102	89	77	64	55	49
830	840	168	155	143	130	117	105	92	79	67	57	50
840	850	171	158	145	133	120	107	95	82	69	58	52
850	860	173	161	148	136	123	110	98	85	72	60	53
860	870	176	164	151	138	126	113	100	88	75	62	55
870	880	179	166	154	141	128	116	103	91	78	65	56
880	890	182	169	157	144	131	119	106	93	81	68	58
890	900	185	172	159	147	134	121	109	96	83	71	59
900	910	187	175	162	150	137	124	112	99	86	74	61
910	920	190	178	165	152	140	127	114	102	89	76	64
920	930	193	180	168	155	142	130	117	105	92	79	67
930	940	196	183	171	158	145	133	120	107	95	82	69
940	950	199	186	173	161	148	135	123	110	97	85	72
950	960	202	189	176	164	151	138	126	113	100	88	75
960	970	205	192	179	166	154	141	128	116	103	90	78
970	980	208	194	182	169	156	144	131	119	106	93	81
980	990	211	197	185	172	159	147	134	121	109	96	83
990	1,000	214	200	187	175	162	149	137	124	111	99	86
1,000	1,010	217	203	190	178	165	152	140	127	114	102	89
1,010	1,020	220	206	193	180	168	155	142	130	117	104	92
1,020	1,030	224	210	196	183	170	158	145	133	120	107	95
1,030	1,040	227	213	199	186	173	161	148	135	123	110	97
1,040	1,050	230	216	202	189	176	163	151	138	125	113	100
1,050	1,060	233	219	205	192	179	166	154	141	128	116	103
1,060	1,070	236	222	208	194	182	169	156	144	131	118	106
1,070	1,080	239	225	211	197	184	172	159	147	134	121	109
1,080	1,090	242	228	214	200	187	175	162	149	137	124	111
1,090	1,100	245	231	217	203	190	177	165	152	139	127	114
1,100	1,110	248	234	220	206	193	180	168	155	142	130	117
1,110	1,120	251	237	223	209	196	183	170	158	145	132	120
1,120	1,130	255	241	227	213	199	186	173	161	148	135	123
1,130	1,140	258	244	230	216	202	189	176	163	151	138	125
1,140	1,150	261	247	233	219	205	191	179	166	153	141	128
1,150	1,160	264	250	236	222	208	194	182	169	156	144	131
1,160	1,170	267	253	239	225	211	197	184	172	159	146	134
1,170	1,180	270	256	242	228	214	200	187	175	162	149	137
1,180	1,190	273	259	245	231	217	203	190	177	165	152	139
1,190	1,200	276	262	248	234	220	206	193	180	167	155	142
1,200	1,210	279	265	251	237	223	209	196	183	170	158	145
1,210	1,220	282	268	254	240	226	212	198	186	173	160	148
1,220	1,230	286	272	258	244	230	216	202	189	176	163	151
1,230	1,240	289	275	261	247	233	219	205	191	179	166	153
1,240	1,250	292	278	264	250	236	222	208	194	181	169	156

TABLE 9.6 **Table for Wage Bracket Method of Withholding**

If the wages are—		And the number of withholding allowances claimed is—										
		(For Wages Paid in 1993)										
At least	But less than	0	1	2	3	4	5	6	7	8	9	10
		The amount of income tax to be withheld is—										
$3,200	$3,240	$406	$376	$347	$317	$288	$259	$229	$200	$171	$141	$112
3,240	3,280	412	382	353	323	294	265	235	206	177	147	118
3,280	3,320	418	388	359	329	300	271	241	212	183	153	124
3,320	3,360	424	394	365	335	306	277	247	218	189	159	130
3,360	3,400	430	400	371	341	312	283	253	224	195	165	136
3,400	3,440	439	406	377	347	318	289	259	230	201	171	142
3,440	3,480	450	412	383	353	324	295	265	236	207	177	148
3,480	3,520	461	418	389	359	330	301	271	242	213	183	154
3,520	3,560	472	424	395	365	336	307	277	248	219	189	160
3,560	3,600	483	430	401	371	342	313	283	254	225	195	166
3,600	3,640	495	440	407	377	348	319	289	260	231	201	172
3,640	3,680	506	451	413	383	354	325	295	266	237	207	178
3,680	3,720	517	462	419	389	360	331	301	272	243	213	184
3,720	3,760	528	473	425	395	366	337	307	278	249	219	190
3,760	3,800	539	485	431	401	372	343	313	284	255	225	196
3,800	3,840	551	496	441	407	378	349	319	290	261	231	202
3,840	3,880	562	507	452	413	384	355	325	296	267	237	208
3,880	3,920	573	518	463	419	390	361	331	302	273	243	214
3,920	3,960	584	529	475	425	396	367	337	308	279	249	220
3,960	4,000	595	541	486	431	402	373	343	314	285	255	226
4,000	4,040	607	552	497	442	408	379	349	320	291	261	232
4,040	4,080	618	563	508	453	414	385	355	326	297	267	238
4,080	4,120	629	574	519	465	420	391	361	332	303	273	244
4,120	4,160	640	585	531	476	426	397	367	338	309	279	250
4,160	4,200	651	597	542	487	432	403	373	344	315	285	256
4,200	4,240	663	608	553	498	443	409	379	350	321	291	262
4,240	4,280	674	619	564	509	455	415	385	356	327	297	268
4,280	4,320	685	630	575	521	466	421	391	362	333	303	274
4,320	4,360	696	641	587	532	477	427	397	368	339	309	280
4,360	4,400	707	653	598	543	488	433	403	374	345	315	286
4,400	4,440	719	664	609	554	499	444	409	380	351	321	292
4,440	4,480	730	675	620	565	511	456	415	386	357	327	298
4,480	4,520	741	686	631	577	522	467	421	392	363	333	304
4,520	4,560	752	697	643	588	533	478	427	398	369	339	310
4,560	4,600	763	709	654	599	544	489	434	404	375	345	316
4,600	4,640	775	720	665	610	555	500	446	410	381	351	322
4,640	4,680	786	731	676	621	567	512	457	416	387	357	328
4,680	4,720	797	742	687	633	578	523	468	422	393	363	334
4,720	4,760	808	753	699	644	589	534	479	428	399	369	340
4,760	4,800	819	765	710	655	600	545	490	436	405	375	346
4,800	4,840	831	776	721	666	611	556	502	447	411	381	352
4,840	4,880	842	787	732	677	623	568	513	458	417	387	358
4,880	4,920	853	798	743	689	634	579	524	469	423	393	364
4,920	4,960	864	809	755	700	645	590	535	480	429	399	370
4,960	5,000	875	821	766	711	656	601	546	492	437	405	376
5,000	5,040	887	832	777	722	667	612	558	503	448	411	382
5,040	5,080	898	843	788	733	679	624	569	514	459	417	388
5,080	5,120	909	854	799	745	690	635	580	525	470	423	394
5,120	5,160	920	865	811	756	701	646	591	536	482	429	400
5,160	5,200	931	877	822	767	712	657	602	548	493	438	406
5,200	5,240	943	888	833	778	723	668	614	559	504	449	412
5,240	5,280	954	899	844	789	735	680	625	570	515	460	418
5,280	5,320	965	910	855	801	746	691	636	581	526	472	424
5,320	5,360	976	921	867	812	757	702	647	592	538	483	430
5,360	5,400	987	933	878	823	768	713	658	604	549	494	439
5,400	5,440	999	944	889	834	779	724	670	615	560	505	450
5,440	5,480	1,010	955	900	845	791	736	681	626	571	516	462
5,480	5,520	1,021	966	911	857	802	747	692	637	582	528	473
5,520	5,560	1,032	977	923	868	813	758	703	648	594	539	484
5,560	5,600	1,043	989	934	879	824	769	714	660	605	550	495
5,600	5,640	1,055	1,000	945	890	835	780	726	671	616	561	506
5,640	5,680	1,066	1,011	956	901	847	792	737	682	627	572	518
5,680	5,720	1,077	1,022	967	913	858	803	748	693	638	584	529
5,720	5,760	1,088	1,033	979	924	869	814	759	704	650	595	540
5,760	5,800	1,099	1,045	990	935	880	825	770	716	661	606	551

Example 4

A married worker earns $694.89 per week and claims three exemptions. Use the percentage method to find the federal income tax deducted from the worker's weekly paycheck.

Strategy

To find the amount deducted:

- Use Table 9.3 to find the withholding allowance.

- Subtract the withholding allowance from the weekly earnings to find the federal earnings.

- Use Table 9.4 to calculate the FIT.

Solution

Withholding allowance = 3(45.19)
                      = 135.57

Federal earnings
    = 694.89 − 135.57
    = 559.32

FIT = 0.15(559.32 − 119)
    = 0.15(440.32)
    = 66.05

The amount deducted for FIT is $66.05.

You Try It 4

A single worker earns $2740 per month and claims one exemption. Use the percentage method to find the federal income tax deducted from the worker's monthly paycheck.

Your strategy

Your solution

Example 5

Use the wage bracket method to find the federal income tax to be deducted from a single worker's weekly paycheck of $743.90 if the worker claims two deductions.

Strategy

To find the amount deducted, use Table 9.5. $743.50 is between $740 and $750. Read the figure in the column headed 2 withholding allowances.

Solution

The amount deducted for FIT is $117.

You Try It 5

Use the wage bracket method to find the federal income tax to be deducted from a married worker's monthly paycheck of $4690 if the worker claims four deductions.

Your strategy

Your solution

Solutions on p. A31

EXERCISE 9.2A

Find the FICA tax to be deducted from the given wage for each of the following four workers. Use the tax rate of 7.65%.

Gross Wage	Accumulated Earnings	FICA Tax
$372.00	$8458.00	1.
$864.72	$20,419.90	2.
$2415.55	$42,675.10	3.
$3015.45	$46,455.25	4.

Find the social security tax, the Medicare tax, and the total FICA tax to be deducted from the given wage for each of the following four workers. Use the social security tax rate of 6.2% and the Medicare tax rate of 1.45%.

Gross Wage	Accumulated Earnings	Social Security Tax	Medicare Tax	Total FICA Tax
$396.00	$9634.00	5.	6.	7.
$785.84	$19,329.80	8.	9.	10.
$1786.55	$36,795.20	11.	12.	13.
$2843.35	$55,435.15	14.	15.	16.

Find the FICA tax to be deducted from the given wage for each of the following four workers by using the Social Security Tax Table (Table 9.1) and the Medicare Tax Table (Table 9.2).

Gross Wage	Accumulated Earnings	Social Security Tax	Medicare Tax	Total FICA Tax
$959.63	$9634.00	17.	18.	19.
$1863.79	$19,329.80	20.	21.	22.
$3172.25	$36,795.20	23.	24.	25.
$4981.98	$53,535.15	26.	27.	28.

1. _____
2. _____
3. _____
4. _____
5. _____
6. _____
7. _____
8. _____
9. _____
10. _____
11. _____
12. _____
13. _____
14. _____
15. _____
16. _____
17. _____
18. _____
19. _____
20. _____
21. _____
22. _____
23. _____
24. _____
25. _____
26. _____
27. _____
28. _____

Solve.

29. An account executive receives a salary of $4810 per month. Determine the amount of the December check that is subject to social security tax.

30. A school administrator receives a monthly salary of $5175. Determine the amount of the December check that is subject to social security tax.

31. Use the Social Security Tax Table and the Medicare Tax Table to find the FICA tax to be withheld from an employee's gross pay of $962.44. The accumulated gross earnings are less than $57,600.

32. Use the Social Security Tax Table and the Medicare Tax Table to find the FICA tax to be withheld from an employee's gross pay of $684.55. The accumulated gross earnings are less than $57,600.

33. Use the Social Security Tax Table and the Medicare Tax Table to find the FICA tax to be withheld from Wilson Brauer's gross pay of $3456.78. His accumulated gross earnings before this paycheck were $55,235.67.

34. Use the Social Security Tax Table and the Medicare Tax Table to find the FICA tax to be withheld from Consuela Marabou's gross pay of $4285.54. Her accumulated gross earnings before this paycheck were $54,521.54.

35. Use the tax rate of 7.65% to find the FICA tax to be withheld from Loretta Wells's gross pay of $732.95. Her accumulated gross earnings are less than $57,600.

36. Use the tax rate of 7.65% to find the FICA tax to be withheld from Emmett Klinger's gross pay of $518.50. His accumulated gross earnings are less than $57,600.

37. An office manager receives a monthly salary of $5675. Use the tax rates of 6.2% for social security and 1.45% for Medicare to find (a) the social security tax and (b) the Medicare tax to be deducted from the February paycheck.

38. Use the tax rates of 6.2% for social security and 1.45% for Medicare to find (a) the social security tax and (b) the Medicare tax to be deducted from an employee's gross pay of $850.75. The accumulated gross earnings are less than $57,600.

39. A marketing manager receives a salary of $4985 per month. Use the tax rates of 6.2% for social security and 1.45% for Medicare to find (a) the social security tax and (b) the Medicare tax to be deducted from the December paycheck.

29. _____

30. _____

31. _____

32. _____

33. _____

34. _____

35. _____

36. _____

37. (a) _____

    (b) _____

38. (a) _____

    (b) _____

39. (a) _____

    (b) _____

40. An operations manager receives a salary of $4850 per month. Use the tax rates of 6.2% for social security and 1.45% for Medicare to find (a) the social security tax and (b) the Medicare tax to be deducted from the December paycheck.

40. (a) _____

    (b) _____

41. _____

42. _____

43. _____

44. _____

45. _____

46. _____

EXERCISE 9.2B

Use the percentage method to find the federal earnings and the federal income tax (FIT) to be withheld from the given wages of each of the following six workers.

47. _____

48. _____

Wages	Payroll Period	Married	Withholding Allowances	Federal Earnings	FIT
$3327	Monthly	Yes	4	41.	42.
$4156.45	Monthly	No	2	43.	44.
$549	Weekly	No	1	45.	46.
$913.37	Weekly	Yes	3	47.	48.
$1307.70	Biweekly	Yes	5	49.	50.
$2083.33	Semimonthly	No	4	51.	52.

49. _____

50. _____

51. _____

52. _____

53. _____

54. _____

55. _____

56. _____

57. _____

58. _____

Use the wage bracket method to find the federal income tax (FIT) to be withheld from the given wages of each of the following six workers.

Wages	Payroll Period	Married	Withholding Allowances	FIT
$3625.00	Monthly	Yes	3	53.
$4271.50	Monthly	Yes	6	54.
$658.47	Weekly	No	0	55.
$802.10	Weekly	No	1	56.
$5318.00	Monthly	Yes	4	57.
$1115.75	Weekly	No	2	58.

Solve.

59. Chris Santiago receives a gross monthly salary of $7166.67. Chris is married and claims two exemptions. Use the percentage method to find (a) the federal earnings and (b) the FIT to be withheld from the monthly paycheck.

60. A heavy-equipment operator receives gross pay of $1424 per week. The operator is single and claims three exemptions. Use the percentage method to find (a) the federal earnings and (b) the FIT to be withheld from the weekly paycheck.

61. Meg Foster earns $2375 biweekly. Meg is married and claims one exemption. Use the percentage method to find the FIT to be withheld from her biweekly paycheck.

62. Jeremy Littlefield earns $3125 semimonthly. Jeremy is married and claims one exemption. Use the percentage method to find the FIT to be withheld from his biweekly paycheck.

63. Saul Levinson earns $947.50 semimonthly. Saul is single and claims two exemptions. Use the percentage method to find the FIT to be withheld from his semimonthly paycheck.

64. Jacqui Mejida earns $947.50 biweekly. Jacqui is single and claims no exemptions. Use the percentage method to find the FIT to be withheld from her semi-monthly paycheck.

65. Use the wage bracket method to find the FIT to be deducted from a single worker's weekly paycheck of $597.40 if the worker claims three exemptions.

66. Use the wage bracket method to find the FIT to be deducted from a married worker's monthly paycheck of $3346 if the worker claims five exemptions.

67. Sidney Banner earns $4210 per month. He is married and claims four exemptions. Find the FIT to be withheld from Sidney's monthly paycheck using (a) the percentage method and (b) the wage bracket method.

68. Larry Parish earns $945 per week. He is single and claims two exemptions. Find the FIT to be withheld from Larry's weekly paycheck using (a) the percentage method and (b) the wage bracket method.

59. (a) _____
    (b) _____
60. (a) _____
    (b) _____
61. _____
62. _____
63. _____
64. _____
65. _____
66. _____
67. (a) _____
    (b) _____
68. (a) _____
    (b) _____

SECTION 9.3 Net Pay

Objective 9.3A *To calculate net pay*

Net pay is the amount of money an employee receives after all deductions from gross earnings have been taken. Some deductions, such as FICA and federal income tax, are mandated by law. Other deductions, such as state income tax and state disability insurance, may be mandated by state law. Other deductions are voluntary—for example, health insurance, life insurance, and savings plans.

Net pay = gross pay − all deductions

An electrician's weekly gross pay is $965. The electrician is married and claims three dependents. Find the net pay after deducting the FICA tax and FIT. Use the tax rate of 7.65% for FICA. Use the percentage method to find the FIT.

- Calculate the FICA tax.

    FICA tax $= 965 \times 0.0765 \approx 73.82$

- Use Tables 9.3 and 9.4 to find the FIT.

    $$\begin{aligned} \text{Federal earnings} &= 965 - (3 \times 45.19) \\ &= 965 - 135.57 \\ &= 829.43 \end{aligned}$$

    $$\begin{aligned} \text{FIT} &= 99.75 + 0.28(829.43 - 784) \\ &= 99.75 + 0.28(45.43) \\ &\approx 99.75 + 12.72 = 112.47 \end{aligned}$$

- Subtract the FICA tax and FIT from the gross pay.

    $$\begin{aligned} \text{Net pay} &= 965 - (73.82 + 112.47) \\ &= 965 - 186.29 \\ &= 778.71 \end{aligned}$$

The net pay is $778.71.

A dental hygienist earns a weekly salary of $720. Deductions are taken for hospitalization ($24.84), life insurance ($18.40), a retirement plan ($50), FICA, and FIT. The hygienist is single and claims one exemption. Find the hygienist's net pay. Use the tax rate of 7.65% for FICA. Use the wage bracket method to find the FIT.

- Find the FICA tax.

    FICA tax $= 720 \times 0.0765 = 55.08$

- Use Table 9.5 to find the FIT.

    FIT $= 124$

- Add the deductions.

    $$\begin{aligned} \text{Deductions} &= 24.84 + 18.40 + 50 + 55.08 + 124 \\ &= 272.32 \end{aligned}$$

- Subtract the deductions from the gross pay.

    Net pay $= 720 - 272.32 = 447.68$

The net pay is $447.68.

Example 1

A single person claiming four exemptions earns $2215 semimonthly. The state income tax is 2.5% of gross earnings, the state disability tax is 1% of gross pay, and $125 is deducted for bonds. Find the net pay. Use the tax rate of 7.65% for FICA. Use the percentage method to find the FIT.

Strategy

To find the net pay:

* Find the FICA tax.

* Find the FIT.

* Find the state income tax.

* Find the state disability tax.

* Add the deductions.

* Subtract the deductions from the gross pay.

Solution

FICA tax = 2215×0.0765
≈ 169.45

Federal earnings = $2215 - (4 \times 97.92)$
$= 2215 - 391.68$
$= 1823.32$

FIT = $130.80 + 0.28(1823.32 - 977)$
$= 130.80 + 0.28(846.32)$
$\approx 130.80 + 236.97$
$= 367.77$

State income tax = 2215×0.025
≈ 55.38

State disability tax = 2215×0.01
$= 22.15$

Deductions
$= 169.45 + 367.77 + 55.38 + 22.15 + 125$
$= 739.75$

Net pay = $2215 - 739.75$
$= 1475.25$

The net pay is $1475.25.

You Try It 1

A married person claiming three exemptions earns $987 biweekly. The state income tax is 3% of gross earnings, union dues are $18.50, and hospitalization insurance is $15.24. Find the net pay. Use the tax rate of 7.65% for FICA. Use the percentage method to find the FIT.

Your strategy

Your solution

Solution on p. A32

Objective 9.3B   *To complete a payroll register*

The accumulated earnings, gross pay, deductions, and net pay of all employees of a company are recorded in a **payroll register** to comply with government regulations. The information for each employee is then transferred for each pay period to the **employee's individual earnings record.** An example of a company's payroll earnings record and an example of an employee's individual earnings record are shown on the next page.

After studying Example 2, work You Try It 2 on page 268.

Example 2

Complete the payroll record for a married teacher with four exemptions. The state withholding tax is 2%, and $50 is deducted for health insurance. Use the FICA tax rate of 7.65%. Use the wage bracket method to find the FIT.

PAY PERIOD	EARNINGS MONTHLY SALARY	DEDUCTIONS					NET PAY
		FICA	FIT	STATE	OTHER	TOTAL	
4/1/94	$3240	_____	_____	_____	_____	_____	_____

Strategy

To complete the payroll record:

- Find the FICA tax.
- Use Table 9.6 to find the FIT.
- Find the state tax.
- Find the total of the deductions.
- Subtract the deductions from the gross pay.

Solution

FICA tax $= 3240 \times 0.0765$
$\qquad = 247.86$

FIT $= 294$

State tax $= 3240 \times 0.02$
$\qquad = 64.80$

Total deductions
$\qquad = 247.86 + 294 + 64.80 + 50$
$\qquad = 656.66$

Net pay $= 3240 - 656.66$
$\qquad = 2583.34$

The net pay is $2583.34.

PAYROLL FOR

	NAME	TOTAL HOURS	BEGINNING CUMULATIVE EARNINGS	EARNINGS			ENDING CUMULATIVE EARNINGS
				REGULAR	OVERTIME	TOTAL	
1	Anaya, Silvano P.	46	6 4 6 0 00	3 2 0 00	7 2 00	3 9 2 00	6 8 5 2 00
2	Brewer, Dewey A.	45	6 7 6 0 00	4 0 0 00	7 5 00	4 7 5 00	7 2 3 5 00
3	Delap, Debbie R.	49	2 3 1 0 0 00	4 0 0 00	1 3 5 00	5 3 5 00	2 3 6 3 5 00
4	Dirksen, Carol B.	40	1 5 8 4 0 00	3 6 0 00	———	3 6 0 00	1 6 2 0 0 00
5	Fife, Arden N.	40	6 1 2 6 00	3 8 4 00	———	3 8 4 00	6 5 1 0 00
6	Garlini, Anthony N.	40	5 4 2 0 0 00	1 2 6 0 00	———	1 2 6 0 00	5 5 4 6 0 00
7	Logan, John C.	52	2 3 3 1 6 00	4 8 0 00	2 4 0 00	7 2 0 00	2 4 0 3 6 00
8	Lucero, Maria D.	40	2 6 4 0 0 00	6 0 0 00	———	6 0 0 00	2 7 0 0 0 00
9	Moser, Donna G.	44	2 1 7 6 2 00	4 4 0 00	6 6 00	5 0 6 00	2 2 2 6 8 00
10	Olson, Lavern H.	45	1 9 3 3 4 00	4 0 0 00	7 5 00	4 7 5 00	1 9 8 0 9 00
11	Pearl, Thomas M.	40	3 7 4 0 0 00	8 5 0 00	———	8 5 0 00	3 8 2 5 0 00
12	Taber, Martha R.	52	2 7 7 1 2 00	4 8 0 00	2 1 6 00	6 9 6 00	2 8 4 0 8 00
13			2 6 8 4 1 0 00	6 3 7 4 00	8 7 9 00	7 2 5 3 00	2 7 5 6 6 3 00

EMPLOYEE'S INDIVIDUAL EARNINGS RECORD

NAME _John Charles Logan_   EMPLOYEE NO. _5_
ADDRESS _6242 Baxter Drive_   SOC. SEC. NO. _543-24-1680_
        Weston, South Dakota 57816   PAY RATE _$12.00_
MALE _X_   FEMALE _____   EQUIVALENT HOURLY RATE _$12.00_
MARRIED _X_   SINGLE _____   DATE TERMINATED _____
PHONE NO. _663-2556_   DATE OF BIRTH _9/19/51_   CLASSIFICATION FOR WORKERS' COMPENSATION INSURANCE _Sales floor_

			HOURS WORKED		EARNINGS				DEDUCTIONS	
LINE NO.	PERIOD ENDED	DATE PAID	REG.	O.T.	REGULAR	OVERTIME	TOTAL	CUMULATIVE EARNINGS	FEDERAL INCOME TAX	STATE INCOME TAX
40	9/3	9/4	40	8	4 8 0 00	1 4 4 00	6 2 4 00	2 1 1 6 2 00	6 5 00	1 3 00
41	9/10	9/11	40	2	4 8 0 00	3 6 00	5 1 6 00	2 1 6 7 8 00	4 8 00	9 6 0
42	9/17	9/18	40	2	4 8 0 00	3 6 00	5 1 6 00	2 2 1 9 4 00	4 8 00	9 6 0
43	9/24	9/25	40	5	4 8 0 00	9 0 00	5 7 0 00	2 2 7 6 4 00	5 7 00	1 1 4 0
44	9/30	10/1	40	4	4 8 0 00	7 2 00	5 5 2 00	2 3 3 1 6 00	5 4 00	1 0 8 0
45	10/7	10/8	40	12	4 8 0 00	2 4 0 00	7 2 0 00	2 4 0 3 6 00	8 0 00	1 6 0 0

WEEK ENDED *October 7, 1994*

| | TAXABLE EARNINGS | | | DEDUCTIONS | | | |
UNEMPL.	SOCIAL SECURITY	MEDICARE	FEDERAL INCOME TAX	STATE INCOME TAX	SOCIAL SECURITY TAX	MEDICARE TAX	MEDICAL INSURANCE
392 00	392 00	392 00	43 00	8 60	24 30	5 68	19 00
240 00	475 00	475 00	55 00	11 00	29 45	6 89	19 00
	535 00	535 00	58 00	11 60	33 17	7 76	23 00
	360 00	360 00	38 00	7 60	22 32	5 22	19 00
384 00	384 00	384 00	41 00	8 20	23 81	5 57	19 00
		1260 00	215 00	43 00		18 27	25 00
	720 00	720 00	80 00	16 00	44 64	10 44	25 00
	600 00	600 00	62 00	12 40	37 20	8 70	23 00
	506 00	506 00	59 00	11 80	31 37	7 34	23 00
	475 00	475 00	55 00	11 00	29 45	6 89	19 00
	850 00	850 00	112 00	22 40	52 70	12 33	25 00
	696 00	696 00	88 00	17 60	43 15	10 09	25 00
1016 00	5993 00	7253 00	906 00	181 20	371 56	105 18	264 00

| | | | | PAYMENTS | | EXPENSE ACCOUNT DEBITED | | |
OTHER		TOTAL		NET AMOUNT	CK. NO.	SALES SALARY EXPENSE	OFFICE SALARY EXPENSE	
		100 58		291 42	931	392 00		1
UW	3 00	124 34		350 66	932	475 00		2
UW	3 00	136 53		398 47	933	535 00		3
		92 14		267 86	934		360 00	4
		97 58		286 42	935	384 00		5
UW	5 00	306 27		953 73	936		1260 00	6
UW	3 00	179 08		540 92	937	720 00		7
UW	3 00	146 30		453 70	938		600 00	8
		132 51		373 49	939	506 00		9
AR	40 00	161 34		313 66	940	475 00		10
UW	4 00	228 43		621 57	941	850 00		11
UW	3 00	186 84		509 16	942	696 00		12
	64 00	1891 94		5361 06		5033 00	2220 00	13

DATE EMPLOYED *2/1/93*
NO. OF EXEMPTIONS *3*
PER HOUR *X*   PER DAY
PER WEEK   PER MONTH

| DEDUCTIONS | | | | | | PAID | |
SOCIAL SECURITY TAX	MEDICARE TAX	MEDICAL INSURANCE	OTHER CODE	OTHER AMOUNT	TOTAL	NET AMOUNT	CK. NO.
38 69	9 05	25 00	UW	3 00	153 74	470 26	877
31 99	7 48	25 00	UW	3 00	125 07	390 93	889
31 99	7 48	25 00	UW	3 00	125 07	390 93	901
35 34	8 27	25 00	UW	3 00	140 01	429 99	913
34 22	8 00	25 00	UW	3 00	135 02	416 98	925
44 64	10 44	25 00	UW	3 00	179 08	540 92	937

You Try It 2

Complete the payroll record for a single employee with three exemptions. The state withholding tax is 2.5%, and $54 is deducted for health insurance. Use the FICA tax rate of 7.65%. Use the wage bracket method to find the FIT.

PAY PERIOD	EARNINGS WEEKLY SALARY	DEDUCTIONS					NET PAY
		FICA	FIT	STATE	OTHER	TOTAL	
7/1/94	$778	____	____	____	____	____	____

Your strategy

Your solution

Solution on p. A32

EXERCISE 9.3A

Solve.

1. Irene Pappas is married, claims three exemptions, and earns $844 per week. Find (a) the FICA tax, (b) the FIT, and (c) her net pay. Use the tax rate of 7.65% for FICA. Use the percentage method to find the FIT.

2. Joseph Abruzzio is married, claims two exemptions, and earns $788 per week. Find (a) the FICA tax, (b) the FIT, and (c) his net pay. Use the tax rate of 7.65% for FICA. Use the percentage method to find the FIT.

3. Will McGillis is single, claims three exemptions, and earns $610 per week. Find (a) the FICA tax, (b) the FIT, and (c) his net pay. Use the tax rate of 7.65% for FICA. Use the wage bracket method to find the FIT.

4. Lila Kedrova is single, claims two exemptions, and earns $750 per week. Find (a) the FICA tax, (b) the FIT, and (c) her net pay. Use the tax rate of 7.65% for FICA. Use the wage bracket method to find the FIT.

5. Garrick Thorpe claims five exemptions, is married, and earns $3650 per month. The state disability tax is 0.5% of the gross pay, and $150 is deducted for an investment account. Find Garrick's net pay. Use the tax rate of 7.65% for FICA. Use the wage bracket method to find the FIT.

6. Yvonne Pleshette claims four exemptions, is married, and earns $4066 per month. The state disability tax is 1% of the gross pay, and $100 is deducted for an investment account. Find Yvonne's net pay. Use the tax rate of 7.65% for FICA. Use the wage bracket method to find the FIT.

7. James Lee is single and claims one exemption. He earns $4450 per month. Union dues of $36 and a hospitalization insurance premium of $49.50 are deducted. Find James's net pay. Use the tax rate of 7.65% for FICA. Use the percentage method to find the FIT.

8. Frances Duffy is single and claims two exemptions. She earns $3940 per month. Union dues of $44 and a hospitalization insurance premium of $64.50 are deducted. Find her net pay. Use the tax rate of 7.65% for FICA. Use the percentage method to find the FIT.

9. Ronald Walker is married and claims three exemptions. He earns $2040 biweekly. The state income tax is 3% of the gross pay, and the deduction for health insurance is $72.13. Find Ronald's net pay. Use the tax rate of 7.65% for FICA. Use the percentage method to find the FIT.

1. (a) _____

  (b) _____

  (c) _____

2. (a) _____

  (b) _____

  (c) _____

3. (a) _____

  (b) _____

  (c) _____

4. (a) _____

  (b) _____

  (c) _____

5. _____

6. _____

7. _____

8. _____

9. _____

10. Kevin Bailey is married, claims three exemptions, and earns $2150 semimonthly. The state income tax is 4% of the gross pay, and the deduction for health insurance is $68.75. Find Kevin's net pay. Use the tax rate of 7.65% for FICA. Use the percentage method to find the FIT.

11. An electrician worked 52 hours during the week. The electrician receives $18.50 per hour, plus time and a half for overtime. The electrician is married and claims three exemptions. Find (a) the gross pay, (b) the FICA tax, (c) the FIT, and (d) the net pay. Use the tax rate of 7.65% for FICA. Use the percentage method to find the FIT.

12. A plumber worked 48 hours during the week. The plumber receives $19.60 per hour, plus time and a half for overtime. The plumber is married and claims four exemptions. Find (a) the gross pay, (b) the FICA tax, (c) the FIT, and (d) the net pay. Use the tax rate of 7.65% for FICA. Use the percentage method to find the FIT.

13. An account executive receives a salary of $250 per week plus 3% of sales. During the past week, the account executive had $37,500 in sales. The account executive is single and claims two exemptions. Find (a) the gross pay, (b) the FICA tax, (c) the FIT, and (d) the net pay. Use the tax rate of 7.65% for FICA. Use the percentage method to find the FIT.

14. An insurance broker receives a salary of $300 per week plus 2% of sales. During the past week, the broker had $52,000 in sales. The insurance broker is single and claims one exemption. Find (a) the gross pay, (b) the FICA tax, (c) the FIT, and (d) the net pay. Use the tax rate of 7.65% for FICA. Use the percentage method to find the FIT.

15. Lori Yusko is single and claims one exemption. She earns $4820 per month. The city income tax is 1.5% of the gross pay. Find Lori's net pay for the month of December. Use the tax rates of 6.2% for social security and 1.45% for Medicare. Use the percentage method to find the FIT.

16. Michael Chan is single and claims three exemptions. He earns $5140 per month. The city income tax is 1.25% of the gross pay. Find Michael's net pay for the month of December. Use the tax rates of 6.2% for social security and 1.45% for Medicare. Use the percentage method to find the FIT.

17. Siobhan McIver is married and claims two exemptions. She earns $4970 per month. The state disability tax is 2% of the gross pay. Find Siobhan's net pay for the month of December. Use the tax rates of 6.2% for social security and 1.45% for Medicare. Use the wage bracket method to find the FIT.

18. Sarah Ontkean is married and claims four exemptions. She earns $5350 per month. The state income tax is 4% of the gross pay. Find Sarah's net pay for the month of November. Use the tax rates of 6.2% for social security and 1.45% for Medicare. Use the wage bracket method to find the FIT.

10. _____

11. (a) _____

    (b) _____

    (c) _____

    (d) _____

12. (a) _____

    (b) _____

    (c) _____

    (d) _____

13. (a) _____

    (b) _____

    (c) _____

    (d) _____

14. (a) _____

    (b) _____

    (c) _____

    (d) _____

15. _____

16. _____

17. _____

18. _____

*E*XERCISE 9.3B

Complete the employee earnings record for the month of March. Use the tax rate of 7.65% for FICA. Use the wage bracket method to find the FIT.

NAME _____		MARRIED _No_		EXEMPTIONS _2_		
ADDRESS _____		REGULAR RATE OF PAY _12.60_		OVERTIME RATE _18.90_		
PAY PERIOD	HOURS	EARNINGS	DEDUCTIONS			NET PAY
			FICA	FIT	TOTAL	
3/5	46	19.	20.	21.	22.	23.
3/12	50	24.	25.	26.	27.	28.
3/19	54	29.	30.	31.	32.	33.
3/26	45	34.	35.	36.	37.	38.

Complete the following earnings record for an employee who receives a salary of $250 per week plus 2% of sales. The employee is married and claims three exemptions. Use the tax rate of 6.2% for FICA/OASI (social security). Use the tax rate of 1.45% for FICA/MED (Medicare). Use the percentage method to find the FIT. The accumulated earnings for the year to date had been $16,300.

PAY PERIOD	SALES	EARNINGS	DEDUCTIONS				NET PAY
			FICA/ OASI	FICA/ MED	FIT	TOTAL	
7/3	8,640	39.	40.	41.	42.	43.	44.
7/10	12,500	45.	46.	47.	48.	49.	50.
7/17	19,250	51.	52.	53.	54.	55.	56.
7/24	24,420	57.	58.	59.	60.	61.	62.

19. _____ 20. _____

21. _____ 22. _____

23. _____ 24. _____

25. _____ 26. _____

27. _____ 28. _____

29. _____ 30. _____

31. _____ 32. _____

33. _____ 34. _____

35. _____ 36. _____

37. _____ 38. _____

39. _____ 40. _____

41. _____ 42. _____

43. _____ 44. _____

45. _____ 46. _____

47. _____ 48. _____

49. _____ 50. _____

51. _____ 52. _____

53. _____ 54. _____

55. _____ 56. _____

57. _____ 58. _____

59. _____ 60. _____

61. _____ 62. _____

Complete the following payroll record for a sales account executive receiving a 3% commission on sales. The executive is married and claims four exemptions. The accumulated earnings at the beginning of August were $42,570. To find the FICA tax, use the tax rates of 6.2% for social security and 1.45% for Medicare. Use the wage bracket method to find the FIT.

| PAY PERIOD | SALES | EARNINGS | DEDUCTIONS | | | NET PAY |
			FICA	FIT	TOTAL	
Aug	142,000	63.	64.	65.	66.	67.
Sept	108,300	68.	69.	70.	71.	72.
Oct	167,500	73.	74.	75.	76.	77.
Nov	132,000	78.	79.	80.	81.	82.

Complete the payroll record for a single employee with three exemptions. The state withholding tax is 2.5% and $150 is deducted for a retirement fund. Use the FICA tax rate of 7.65%. Use the percentage method to find the FIT.

| PAY PERIOD | EARNINGS | DEDUCTIONS | | | | | NET PAY |
		FICA	FIT	STATE	OTHER	TOTAL	
June	4873	83.	84.	85.	86.	87.	88.

63. _____

64. _____

65. _____

66. _____

67. _____

68. _____

69. _____

70. _____

71. _____

72. _____

73. _____

74. _____

75. _____

76. _____

77. _____

78. _____

79. _____

80. _____

81. _____

82. _____

83. _____

84. _____

85. _____

86. _____

87. _____

88. _____

CALCULATORS

+/− **Key**   The +/− key on a hand-held calculator can simplify the calculation of such items as taxable salary, federal earnings, and net pay. Using this key eliminates the need to re-enter numbers.

In Example 3 on page 252, Shelly Egan earns a gross monthly salary of $5325. The amount of the November paycheck that is subject to social security tax can be calculated as follows:

Enter	Display	Comments
10 × 5325 =	53250	Calculate the gross salary earned thus far during the year.
+/−	−53250	The +/− key changes the sign in front of the number in the display.
+ 57600 =	4350	Here 51,350 was subtracted from 57,600.

The taxable salary for social security is $4350.

In Example 4 on page 258, the worker earns $694.89 per week and claims three exemptions. The federal earnings of the worker can be calculated as follows:

Enter	Display	Comments
3 × 45.19 =	135.57	Calculate the withholding allowance.
+/−	−135.57	A minus sign is now in front of the number.
+ 694.89 =	559.32	Here 135.57 was subtracted from 694.89.

The federal earnings are $559.32.

Solve.

1. Mohammed Arau earns a gross salary of $4825 per month. Find the amount of his December salary that is subject to social security tax.

1. _____

2. _____

3. _____

2. Anita Pertana is married and claims five exemptions. She earns $1575 per week. Use the percentage method to find the federal earnings.

3. Derrick Jackson earns a monthly salary of $3540 and has deductions of $270.81, $361, and $75. Find Derrick's net pay. (*Hint:* Find the sum of the deductions, use the +/− key, and then add the monthly salary.)

Store/Recall Keys   Complete the following Employer's Payroll Tax Register. Use a FICA tax rate of 7.65%. The State Unemployment Tax (S.U.T.A.) rate is 5.4%, and the Federal Unemployment Tax (F.U.T.A.) rate is 0.8%.

Solve the problems vertically by calculating all the FICA taxes, then all the S.U.T.A. taxes, and then all the F.U.T.A. taxes. For each tax rate, change the percent to a decimal and store the decimal in the calculator's memory. Round answers to the nearest cent.

Employer's Payroll Tax Register **January 31, 1993**				
EMPLOYEE	TAXABLE WAGES	FICA (7.65%)	S.U.T.A. (5.4%)	F.U.T.A. (0.8%)
Abrahms	$869.86	4.	14.	24.
Bavis	$732.41	5.	15.	25.
Chavez	$791.54	6.	16.	26.
Deane	$828.47	7.	17.	27.
Evans	$679.85	8.	18.	28.
Nguyen	$747.82	9.	19.	29.
Grant	$693.90	10.	20.	30.
Hopkins	$743.60	11.	21.	31.
Nevarez	$921.75	12.	22.	32.
Keene	$824.95	13.	23.	33.

4. _____

5. _____

6. _____

7. _____

8. _____

9. _____

10. _____

11. _____

12. _____

13. _____

14. _____

15. _____

16. _____

17. _____

18. _____

19. _____

20. _____

21. _____

22. _____

23. _____

24. _____

25. _____

26. _____

27. _____

28. _____

29. _____

30. _____

31. _____

32. _____

33. _____

BUSINESS CASE STUDY

Kite Manufacturing Company

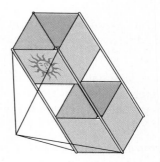

The company was started by Mike, a 20-year-old college student, who loves kites and designed and built his own kites as a kid. Friends began asking him to build kites for them, and before long he outgrew his father's garage and moved into a commercial building to manufacture and sell his kites. The business is now six years old and is enjoying steady growth. Mike's kites are distributed and sold in specialty toy stores.

Currently there are 35 employees:

Mike	President
Sue	Vice President of Production and Distribution
Larry	Vice President of Marketing and Personnel
3	Managers
4	Sales representatives
25	Production workers

Mike, Sue, and Larry are paid a monthly salary. The managers are paid a semimonthly salary. The sales representatives are paid a semimonthly salary plus commission. The production workers are paid weekly; they receive an hourly wage plus time and a half for working more than 40 hours per week.

Payroll information on some of the employees follows.

Employee	Status	Allowances	Salary or Wage
Mike	Married	3	$2400
Sue	Married	1	$1800
Larry	Single	0	$1800
John (manager)	Married	4	$1600
Carol (manager)	Single	2	$1550
Tony (sales)	Single	1	$500
Mary (sales)	Married	2	$500
Randy (sales)	Single	0	$500
Barbara (production)	Married	3	$6 per hour
Todd (production)	Married	5	$7.50 per hour
Allen (production)	Single	1	$6.25 per hour

For the period May 1–May 15, Tony had sales of $8525, Mary had sales of $9730, and Randy had sales of $6990. The graduated commission scale is as shown at the right.

$0–$5000	5%
$5001–$7500	6%
$7501–$10,000	7%
$10,000 and above	8%

The daily hours worked for three of the production workers the first week of May were

	Monday	Tuesday	Wednesday	Thursday	Friday
Barbara	8	8	9	8	10
Todd	8	8	8	8	8
Allen	7	9	8	8	10

1. Calculate the gross pay for (a) Barbara, Todd, and Allen for the week of May 1; (b) John, Carol, Tony, Mary, and Randy for May 15; and (c) John, Carol, Mike, Sue, and Larry for May 30.
2. Calculate FICA and the federal income tax deduction for each employee in Exercise 1. Use a tax rate of 7.65%. Use the percentage method to find the FIT.
3. Calculate the net pay (gross pay minus FICA minus FIT) for each employee in Exercise 1.

CHAPTER SUMMARY

Key Words   **Gross earnings**, or **gross pay**, is the total amount an employee has earned. **Net earnings**, or **net pay**, is the amount of money an employee receives after all deductions from gross earnings have been made. (Objectives 9.1A and 9.3A)

A **commission** is a percent of sales, called a **commission rate**, or an amount per item sold. Under a **variable** or **graduated commission scale**, different commission rates are used for different levels of sales. (Objective 9.1B)

Employees on a **piecework** plan are paid on the basis of the number of items they produce that pass inspection. A piecework rate structure that depends on a tiered scale is called a **differential piece rate** (as the number of items produced increases, the rate per item paid to the employees also increases). (Objective 9.1C)

Deductions are amounts of money that are subtracted from gross pay. Mandatory deductions include **FICA** (Federal Insurance Contributions Act), or social security, and federal income tax (**FIT**). (Objectives 9.2A and 9.2B)

The number of exemptions, or dependents, claimed by an employee on a W-4 form is referred to as the number of **withholding allowances**. The withholding allowance is subtracted from gross pay in order to find the **federal earnings**, which is the amount subject to federal withholding. (Objective 9.2B)

To determine the amount of federal withholding tax to be deducted from a paycheck, one of two methods is used: the **percentage method** or the **wage bracket method**. Both methods rely on tables supplied by the federal government. (Objective 9.2B)

The accumulated earnings, gross pay, deductions, and net pay of all employees of a company are recorded in a **payroll register**. The information for each employee is then transferred for each pay period to the **employee's individual earnings record**. (Objective 9.3B)

Essential Rules   **Gross pay for an employee paid an hourly wage**
      = number of hours worked × rate per hour (Objective 9.1A)

Gross pay for employees paid for overtime
      = regular pay + overtime pay (Objective 9.1A)

Payroll Periods:
      Monthly—12 paychecks per year
      Semimonthly (twice a month)—24 paychecks per year
      Weekly—52 paychecks per year
      Biweekly (every 2 weeks)—26 paychecks per year (Objective 9.1A)

Gross pay per period = annual salary ÷ number of pay periods in a year (Objective 9.1A)

Gross pay (commission only) = total sales × commission rate
Gross pay (combination salary and commission) = salary + commission (Objective 9.1B)

Gross pay for piecework = rate per item × number of items produced that pass inspection (Objective 9.1C)

FICA rate of taxation in 1993 = 7.65%: 6.2% of earnings up to $57,600 for the social security tax, and 1.45% of earnings up to $135,000 for the Medicare tax (Objective 9.2A)

Net pay = gross pay − all deductions (Objective 9.3A)

REVIEW / TEST

1. A clerk worked $6\frac{1}{2}$ hours on Monday, 8 hours on Tuesday, $7\frac{1}{4}$ hours on Wednesday, and 12 hours on Thursday. The clerk receives $9.50 per hour. Find the clerk's gross pay for the week.

2. A carpenter worked 50 hours during this week. The carpenter receives $18.20 per hour for 40 hours and time and a half for overtime. Find the carpenter's gross pay for this week.

3. A real estate agent receives a commission of 3% of sales. Find the agent's gross pay for selling a house for $132,500.

4. A clerk at a jewelry store earns a wage of $8.65 per hour and a 2% commission on sales. The clerk sold $6000 of merchandise and worked 35 hours during the past week. Find the clerk's gross pay.

5. An orange picker receives $.95 for picking a box of oranges. Find the weekly pay when the worker picks 324 boxes of oranges during the week.

6. A technician earns $6.50 per hour and a piecework rate of $.80 per unit. The technician worked 40 hours and produced 400 units this week. Find the technician's gross pay for the week.

7. A sales representative is allowed a drawing account of $750 per month and receives a monthly commission of 3% of the first $20,000 of sales and 4% of all sales in excess of $20,000. Find the amount due the sales representative for a month in which the representative's drawings were $700 and sales totaled $35,000.

8. A welder receives pay on a variable piecework basis. The welder receives $.70 for every weld up to 1200 and $.90 for all welds over 1200. Find the gross pay for a week in which the welder completed 1400 welds.

1. _____

2. _____

3. _____

4. _____

5. _____

6. _____

7. _____

8. _____

9. Find the FICA tax to be deducted from gross earnings of $814.56. Use the tax rate of 7.65%.

10. Find (a) the social security tax and (b) the Medicare tax to be deducted from gross earnings of $4328. The accumulated gross earnings are $53,892. Use the social security tax rate of 6.2% and the Medicare tax rate of 1.45%.

11. Use the wage bracket method to find the FIT deduction from a married worker's monthly earnings of $4368. The worker claims two exemptions.

12. Use the percentage method to find the FIT deduction from a married worker's biweekly paycheck of $2236. The worker claims three exemptions.

13. Buford Fire Cloud is single and claims two exemptions. He earns $862 per week. Find Buford's net pay. Use the tax rate of 7.65% for FICA. Use the wage bracket method to find the FIT.

14. Karen Hill is married, claims four exemptions, and earns $942 per week. The state income tax is 3.5% of the gross pay, and $58.40 is deducted for health insurance. Find Karen's net pay. Use the tax rate of 7.65% for FICA. Use the percentage method to find the FIT.

9. _____

10. (a) _____

    (b) _____

11. _____

12. _____

13. _____

14. _____

15. _____

16. _____

17. _____

18. _____

19. _____

20. _____

Complete the payroll record for a married employee with two exemptions. The state withholding tax is 1.5%, and $200 is deducted for savings. Use the tax rate of 7.65% for FICA. Use the percentage method to find the FIT.

PAY PERIOD	WEEKLY EARNINGS	DEDUCTIONS					NET PAY
		FICA	FIT	STATE	OTHER	TOTAL	
3/7	1,027	15.	16.	17.	18.	19.	20.

10

Simple and Compound Interest

OBJECTIVES

10.1A   To compute simple interest

10.1B   To compute the maturity value and the maturity date of a note

10.2A   To calculate the discount and proceeds of a note

10.3A   To find the future value of compound interest

10.3B   To find the present value of compound interest

10.3C   To calculate effective interest rates

*I*N THE FIRST two sections of this chapter, we will be studying simple interest and its application to promissory notes and simple discount notes. In the third section, the concept of compound interest is developed.

An understanding of future value and present value in connection with compound interest will enable you to answer such questions as

If I put $2000 in a certificate of deposit today, what will it be worth in 6 months?

How much interest will the money in my savings account earn over the next year?

How much money do I need to invest today in order to have $20,000 for a down payment on a house 4 years from now?

How can I determine which bank is offering the highest rates on savings deposits?

The **Rule of 72**, which involves a very simple calculation, provides a quick and easy method of estimating how long it will take for an investment to double in value: divide the number 72 by the interest rate. For example, an investment earning 9% interest will double in value in approximately $72 \div 9 = 8$ years. How long will it take an investment earning 6% interest to double?[1]

The application of compound interest to investments is really quite amazing. Let's assume an investment account earns 9% interest compounded monthly. How much money do you think you would have to deposit in this account every month, starting at age 25, in order to have $1,000,000 when you retire at the age of 65? The answer is less than $215 per month. How much have you deposited into the account? There are $40 \times 12 = 480$ months in the 40-year period from age 25 to 65. Let's assume you have deposited $215 per month for 480 months.

$$\$215 \times 480 = \$103,200$$

The $103,200 you have saved has grown into over $1 million!

An understanding of the concepts presented in this chapter is crucial to the study of annuities and of business and consumer loans. It also has application to topics studied in earlier chapters of this text. For example, interest is a part of banking (the topic of Chapter 4); NOW accounts, savings accounts, and bank certificates of deposits earn interest. And in the Business Case Study at the end of this chapter, the concept of interest is applied to cash discounts, which we studied in Chapter 7.

[1]12 years

SECTION 10.1   Simple Interest and Maturity Value

Objective 10.1A   *To compute simple interest*

When money is deposited in a bank account, the bank pays the depositor for the privilege of using that money. When money is borrowed from a bank, the borrower pays the bank for the privilege of using that money. The original amount deposited or borrowed is called the **principal.** The amount paid for the privilege of using the money is called **interest.** The amount of interest to be paid is usually computed as a percent of the principal. The percent used to determine the amount of interest to be paid is the **interest rate.**

Interest computed on the original principal is called **simple interest.** Simple interest is given by the formula

Simple Interest	I = simple interest
$I = Prt$	P = principal
	r = annual interest rate as a decimal
	t = time of the loan in years

In the simple interest formula, the interest rate is an annual rate. Therefore, when the time period of a loan is given in days or months, the time of the loan must be converted to years. To convert units of time from days or months to years, we use the following formulas.

Time in months	Time in days (exact method)	Time in days (ordinary method)
$t = \dfrac{\text{number of months}}{12}$	$t = \dfrac{\text{number of days}}{365}$	$t = \dfrac{\text{number of days}}{360}$

Note that when a loan is in terms of days, the time may be calculated by using either the exact method or the ordinary method. The ordinary method is sometimes referred to as the banker's method, because this method is used by many banks.

Find the simple interest on a 45-day loan of $2500 at an annual interest rate of 10.5% by using (a) the exact method and (b) the ordinary method.

♦ Use the simple interest formula.
$P = 2500$, $r = 10.5\% = 0.105$ $I = Prt$

(a) $t = \dfrac{45}{365}$ $I = 2500(0.105)\left(\dfrac{45}{365}\right) \approx 32.36$

(b) $t = \dfrac{45}{360}$ $I = 2500(0.105)\left(\dfrac{45}{360}\right) \approx 32.81$

(a) The exact method yields interest of $32.36. (b) The ordinary method yields interest of $32.81. Note that the interest is greater when the ordinary method is used.

Unless otherwise stated, the ordinary method will be used in this text.

In the previous example, the simple interest formula was solved for I. In the following examples, the simple interest formula is solved for r, for P, and for t.

The simple interest charged on a 6-month loan of \$3000 is \$150. Find the simple interest rate.

* Solve the simple interest formula for r.

$I = 150, P = 3000, t = \dfrac{6}{12}$

$$I = Prt$$
$$150 = 3000(r)\left(\dfrac{6}{12}\right)$$
$$150 = 1500r$$
$$\dfrac{150}{1500} = \dfrac{1500r}{1500}$$

* Write the answer as a decimal.

$$0.1 = r$$

The simple interest rate is 10%.

The simple interest rate on a 9-month loan is 9%. The simple interest charged is \$270. Find the loan amount.

* Solve the simple interest formula for P.

$I = 270, r = 9\% = 0.09, t = \dfrac{9}{12}$

$$I = Prt$$
$$270 = P(0.09)\left(\dfrac{9}{12}\right)$$
$$270 = P(0.0675)$$
$$\dfrac{270}{0.0675} = \dfrac{P(0.0675)}{0.0675}$$
$$4000 = P$$

The loan amount is \$4000.

The simple interest charged on a \$5000 loan is \$100. The simple interest rate is 8%. Find the time of the loan (a) in months and (b) in days.

* Solve the simple interest formula for t.

$I = 100, P = 5000, r = 8\% = 0.08$

$$I = Prt$$
$$100 = 5000(0.08)t$$
$$100 = 400t$$
$$\dfrac{100}{400} = \dfrac{400t}{400}$$

The time is 1/4 of a year.

$$0.25 = t$$

* Convert the time of the loan to months. Multiply by the number of months in one year (12).

$$0.25(12) = 3$$

* Convert the time of the loan to days. Multiply by the number of days in one year when using the ordinary method (360).

$$0.25(360) = 90$$

The time of the loan is (a) 3 months or (b) 90 days.

Example 1

Find the simple interest on a 6-month loan of $10,000 at an annual interest rate of 7.75%.

Strategy

To find the simple interest, use the simple interest formula $I = Prt$.

$P = 10,000, r = 0.0775, t = \dfrac{6}{12}$

Solution

$I = Prt$

$\quad = 10,000(0.0775)\left(\dfrac{6}{12}\right)$

$\quad = 387.5$

The interest on the loan is $387.50.

You Try It 1

Find the simple interest on a 3-month loan of $6000 at an annual interest rate of 8.4%.

Your strategy

Your solution

Example 2

The simple interest on a 45-day loan of $5200 is $55.25. Find the simple interest rate.

Strategy

To find the simple interest rate, solve the simple interest formula for r.

$I = 55.25, P = 5200, t = \dfrac{45}{360}$

Solution

$\quad I = Prt$

$55.25 = 5200(r)\left(\dfrac{45}{360}\right)$

$55.25 = 650r$

$0.085 = r$

The simple interest rate is 8.5%.

You Try It 2

The simple interest on a 75-day loan of $3000 is $57.50. Find the simple interest rate.

Your strategy

Your solution

Solutions on p. A33

Objective 10.1B   *To compute the maturity value and the maturity date of a note*

A **promissory note,** or simply a **note,** is a contract signed by a borrower for a loan. An important feature of a promissory note is that it is *negotiable;* it can be sold to another person or company.

The principal of a note is called the **face value** of the note. The **maturity value** of a loan or note is the sum of the principal and the interest. The maturity value is the total amount of principal and interest that must be repaid to the lender.

Maturity Value M = maturity value
$M = P + I$ P = principal, or face value
    I = interest earned

Find the maturity value of a simple interest, 6-month promissory note of $8000 at an annual interest rate of 9%.

◆ Find the simple interest. $I = Prt$

$P = 8000, r = 0.09, t = \dfrac{6}{12}$ $= 8000(0.09)\left(\dfrac{6}{12}\right)$

    $= 360$

◆ The maturity value is the prin- $M = P + I$
cipal plus the simple interest. $= 8000 + 360$
    $= 8360$

The maturity value is $8360.

The **maturity date,** or **due date,** of a loan is the date on which the loan and interest must be completely repaid. If a loan is given in years, the maturity date is the corresponding day in the year in which the loan is due. For example, a 2-year loan dated March 12, 1993 has a maturity date of March 12, 1995.

If a loan is given in months, then the maturity date is the corresponding day in the month in which the loan is due. A 3-month loan dated March 15 has a maturity date of June 15. However, a 3-month loan dated November 30 is due on February 28 (or February 29 during a leap year).

For a loan given in days, the maturity date is that number of days after the date of the loan. To find the maturity date of a 60-day loan dated March 28, a calendar could be used to count 60 days past March 28. An alternative method uses the Day of the Year Table shown on the next page.

◆ Locate March 28 in the table. The   87
day of the year appears where   + 60
the Day 28 row and the March   147
column intersect (87). Add the
number of days in the loan to
the day of the year.

◆ Use the table to find the date   Day 147 is May 27.
that corresponds to this day of
the year.

The maturity date of the loan is May 27.

The length of time between the date a note is written and the date the note is due is called the **term** of the note.

DAY OF THE YEAR TABLE

DAYS IN EACH MONTH

		31	28	31	30	31	30	31	31	30	31	30	31
		JAN	FEB	MAR	APR	MAY	JUN	JUL	AUG	SEP	OCT	NOV	DEC
DAY	1	1	32	60	91	121	152	182	213	244	274	305	335
DAY	2	2	33	61	92	122	153	183	214	245	275	306	336
DAY	3	3	34	62	93	123	154	184	215	246	276	307	337
DAY	4	4	35	63	94	124	155	185	216	247	277	308	338
DAY	5	5	36	64	95	125	156	186	217	248	278	309	339
DAY	6	6	37	65	96	126	157	187	218	249	279	310	340
DAY	7	7	38	66	97	127	158	188	219	250	280	311	341
DAY	8	8	39	67	98	128	159	189	220	251	281	312	342
DAY	9	9	40	68	99	129	160	190	221	252	282	313	343
DAY	10	10	41	69	100	130	161	191	222	253	283	314	344
DAY	11	11	42	70	101	131	162	192	223	254	284	315	345
DAY	12	12	43	71	102	132	163	193	224	255	285	316	346
DAY	13	13	44	72	103	133	164	194	225	256	286	317	347
DAY	14	14	45	73	104	134	165	195	226	257	287	318	348
DAY	15	15	46	74	105	135	166	196	227	258	288	319	349
DAY	16	16	47	75	106	136	167	197	228	259	289	320	350
DAY	17	17	48	76	107	137	168	198	229	260	290	321	351
DAY	18	18	49	77	108	138	169	199	230	261	291	322	352
DAY	19	19	50	78	109	139	170	200	231	262	292	323	353
DAY	20	20	51	79	110	140	171	201	232	263	293	324	354
DAY	21	21	52	80	111	141	172	202	233	264	294	325	355
DAY	22	22	53	81	112	142	173	203	234	265	295	326	356
DAY	23	23	54	82	113	143	174	204	235	266	296	327	357
DAY	24	24	55	83	114	144	175	205	236	267	297	328	358
DAY	25	25	56	84	115	145	176	206	237	268	298	329	359
DAY	26	26	57	85	116	146	177	207	238	269	299	330	360
DAY	27	27	58	86	117	147	178	208	239	270	300	331	361
DAY	28	28	59	87	118	148	179	209	240	271	301	332	362
DAY	29	29	0	88	119	149	180	210	241	272	302	333	363
DAY	30	30	0	89	120	150	181	211	242	273	303	334	364
DAY	31	31	0	90	0	151	0	212	243	0	304	0	365

ADD 1 DAY FOR LEAP YEAR IF FEBRUARY 29 FALLS BETWEEN THE TWO DATES

Find the due date of a 90-day loan dated October 15.

♦ From the table, October 15 is Day 288.
Add the number of days in the loan to
the day of the year.

$$\begin{array}{r} 288 \\ +\ 90 \\ \hline 378 \end{array}$$

♦ 378 is greater than 365. The due date is
in the following year. Subtract the
number of days in one year from 378.

$$\begin{array}{r} 378 \\ -\ 365 \\ \hline 13 \end{array}$$

♦ Use the table to find the date that
corresponds to Day 13.

Day 13 is January 13.

The due date of the loan is January 13.

The due date of the loan in the example above could also be found by calculating 90 days from October 15.

♦ 90 days is about 3 months.
3 months from Oct. 15 is Jan. 15.
Count the days from Oct. 15 to Jan. 15.

31 days in Oct.
30 days in Nov.
$$\begin{array}{r} +\ 31 \text{ days in Dec.} \\ \hline 92 \text{ days} \end{array}$$

♦ Subtract 90 from the sum. Oct. 15 to
Jan. 15 is 2 days more than 90 days.

$$\begin{array}{r} -\ 90 \text{ days} \\ \hline 2 \text{ days} \end{array}$$

♦ Subtract 2 days from Jan. 15.

$15 - 2 = 13$

The due date of the loan is January 13.

Example 3

On February 11, a construction company signs a $50,000, 9% simple interest note that is due in 60 days. The year is not a leap year. Find the maturity date and the maturity value of the note.

Strategy

To find the maturity date:

* Find the date 60 days from February 11. (The Day of the Year Table is used in the solution that follows.)

To find the maturity value:

* Find the simple interest by solving the simple interest formula for I.

  $P = 50,000, r = 9\% = 0.09, t = \dfrac{60}{360}$

* Use the maturity value formula $M = P + I$.

Solution

February 11 is Day 42.

$42 + 60 = 102$

Day 102 is April 12.

The maturity date of the note is April 12.

$I = Prt$

$= 50,000(0.09)\left(\dfrac{60}{360}\right)$

$= 750$

$M = P + I$
  $= 50,000 + 750$
  $= 50,750$

The maturity value of the note is $50,750.

You Try It 3

A simple interest, 120-day promissory note of $7000 is signed on May 4. The note has an annual interest rate of 10.4%. Find the maturity date and the maturity value of the note.

Your strategy

Your solution

Solution on p. A33

*E*XERCISE 10.1A

For Exercises 1–12, complete the table.

Interest	Principal	Rate	Time
1.	$16,000	6.5%	8 months
2.	$15,000	8.25%	9 months
3.	$10,000	6.9%	120 days
4.	$17,500	8.4%	60 days
$750	5.	7.5%	3 months
$148.75	6.	8.5%	6 months
$280	$12,000	7.	4 months
$1350	$25,000	8.	9 months
$770	$27,500	8.4%	9. (in months)
$326.25	$9000	7.25%	10. (in months)
$665	$35,000	7.6%	11. (in days)
$229.50	$8500	8.1%	12. (in days)

1. _____

2. _____

3. _____

4. _____

5. _____

6. _____

7. _____

8. _____

9. _____

10. _____

11. _____

12. _____

13. _____

14. _____

Solve.

13. Kevin Moore took out a 120-day loan of $25,000 at an annual interest rate of 9.4%. Find the simple interest due on the loan.

14. Leonore Rishkin took out a 75-day loan of $8000 at an annual interest rate of 8.5%. Find the simple interest due on the loan.

15. Ali Winnisquam borrowed $1500 at an annual interest rate of 11.2%. The loan was for 9 months. Find the simple interest due on the loan.

16. Gregor Andre borrowed $4000 at an annual interest rate of 9.8%. The loan was for 4 months. Find the simple interest due on the loan.

17. What is the simple interest earned on a 75-day loan of $4500 at an annual interest rate of 9%?

18. What is the simple interest earned on a 20-day loan of $20,000 at an annual interest rate of 7.4%?

19. You take out a 3-month loan of $7500 at an annual interest rate of 7.8%. What is the simple interest due on the loan?

20. You take out a 4-month loan of $14,000 at an annual interest rate of 8.6%. What is the simple interest due on the loan?

21. Reggie Means took out a $12,000 loan that earned $462 in interest in 6 months. What was the simple interest rate on Reggie's loan?

22. Kim Himottu took out a $1800 loan that earned $148.50 in interest in 9 months. What was the simple interest rate on Kim's loan?

23. A $24,000 loan earned $450 in interest in 90 days. Find the annual simple interest rate.

24. An $8000 loan earned $136 in interest in 60 days. Find the annual simple interest rate.

25. The simple interest rate on Sean McElduff's 6-month loan was 8%. The simple interest charged was $375. What amount did Sean borrow?

15. _____

16. _____

17. _____

18. _____

19. _____

20. _____

21. _____

22. _____

23. _____

24. _____

25. _____

26. The simple interest rate on Teresa Puelo's 9-month loan was 7.5%. The simple interest charged was $225. What amount did Teresa borrow?

27. The simple interest rate on Lois Maxwell's 90-day loan was 6.4%. What amount did Lois borrow if the simple interest charged was $96?

28. The simple interest rate on Carl Easton's 120-day loan was 8.7%. What amount did Carl borrow if the simple interest charged was $348?

29. Russell Burton borrows $10,000. The simple interest charged is $375, and the simple interest rate is 7.5%. Find the time of Russell's loan (a) in months and (b) in days.

30. Angela Lopez borrows $30,000. The simple interest charged is $510, and the simple interest rate is 6.8%. Find the time of Angela's loan (a) in months and (b) in days.

31. The simple interest rate on Denise Cooper's $12,000 loan is 9.3%. The simple interest on the loan is $372. Find the time of Denise's loan (a) in months and (b) in days.

32. The simple interest rate on Jack Bates's $16,000 loan is 6.9%. The simple interest on the loan is $828. Find the time of Jack's loan (a) in months and (b) in days.

26. _____

27. _____

28. _____

29. (a) _____

    (b) _____

30. (a) _____

    (b) _____

31. (a) _____

    (b) _____

32. (a) _____

    (b) _____

33. _____

34. _____

35. _____

EXERCISE 10.1B

Solve.

33. Find the maturity date of a 6-month note dated August 3.

34. Find the maturity date of a 3-month note dated May 14.

35. A 90-day note is dated July 10. Find the maturity date of the loan.

36. A 120-day note is dated April 4. Find the maturity date of the loan.

37. A 15-day note is dated March 20. Find the due date of the loan.

38. A 45-day note is dated December 8. Find the due date of the loan.

39. Mae Jackson signed a simple interest, 90-day promissory note of $10,000 at an annual interest rate of 8%. Find the maturity value of the note.

40. Dale Matthews signed a simple interest, 30-day promissory note of $25,000 at an annual interest rate of 7.5%. Find the maturity value of the note.

41. The owner of a toy store signs a $40,000, 8.5% simple interest note that is due in 60 days. Find the maturity value of the note.

42. The manager of an automobile dealership signs a $90,000, 9.5% simple interest note that is due in 9 months. Find the maturity value of the note.

43. On September 18, a plumbing store owner signs a 45-day, 8.5% simple interest note of $12,000. Find (a) the due date of the note and (b) the maturity value of the note.

44. On February 16, a record store manager signs a 120-day, 7.5% simple interest note of $750. The year is not a leap year. Find (a) the due date of the note and (b) the maturity value of the note.

45. The manager of a jewelry store signs a 90-day, 9.25% simple interest note of $20,000 on June 14. Find (a) the maturity date of the note and (b) the maturity value of the note.

46. The owner of a dish television antenna store signs a 45-day, 8.75% simple interest note of $15,000 on July 20. Find (a) the maturity date of the note and (b) the maturity value of the note.

36. _____

37. _____

38. _____

39. _____

40. _____

41. _____

42. _____

43. (a) _____

    (b) _____

44. (a) _____

    (b) _____

45. (a) _____

    (b) _____

46. (a) _____

    (b) _____

SECTION 10.2 | Discount Notes

Objective 10.2A   *To calculate the discount and proceeds of a note*

In the last section, it was stated that the principal of a note is the face value and that the maturity value is the sum of the principal and the interest: $M = P + I$. For a simple interest note, the borrower receives the face value (principal) of the note and agrees to repay the maturity value of the note.

However, when a borrower signs a **simple discount note,** the lender first calculates the interest due on the face value of the note and then subtracts that amount from the face value. The borrower receives the face value minus the interest. This amount is called the **proceeds** of the note. In effect, the borrower pays the interest at the beginning of the loan rather than at the end.

The maturity value of a discount note is the sum of the proceeds and the interest. Therefore, for a discount note, the face value and the maturity value are the same.

We can summarize this discussion by stating the following about a discount note:

Calculate the interest due on the face value (principal) of the note.
Subtract the interest due from the face value. This is the proceeds.
The borrower receives the proceeds.
The face value = the maturity value = the proceeds + the interest.
The borrower repays the face value when the note is due.

The interest due on the face value of a simple discount note is calculated by using the simple interest formula, $I = Prt$. In the case of a simple discount note, the interest is called the **discount** (D), the principal is the face value or maturity value (M), and the rate is the **discount rate** (d).

Simple Discount Notes

$D = Mdt$ D = discount
$p = M - D$ M = maturity value or face value
                     d = discount rate
                     t = time in years
                     p = proceeds

A note that has a face value of $5000 and is due in 60 days is discounted 8.4%. Find the discount.

◆ Use the simple discount formula, $D = Mdt$. $D = Mdt$
$M = 5000$, $d = 8.4\% = 0.084$, $t = \dfrac{60}{360}$ $= 5000(0.084)\left(\dfrac{60}{360}\right)$

                                                               $= 70$

The discount is $70.

This means that the simple interest due on the face value of the note is $70.

A bank discounts a 9% simple interest note that has a face value of $1500 and is due in 4 months. Find the proceeds of the note.

- Use the simple discount formula to find the discount.

  $M = 1500$, $d = 9\% = 0.09$, $t = \dfrac{4}{12}$

  $$D = Mdt$$
  $$= 1500(0.09)\left(\dfrac{4}{12}\right)$$
  $$= 45$$

- Use the proceeds formula to find the proceeds.
  $M = 1500$, $D = 45$

  $$p = M - D$$
  $$= 1500 - 45$$
  $$= 1455$$

The proceeds are $1455.

This is the face value minus the interest due. It is what the borrower receives on the note.

The discount on a $6000, 90-day simple discount note is $165. Find the discount rate.

- Solve the simple discount formula for d.

  $D = 165$, $M = 6000$, $t = \dfrac{90}{360}$

  $$D = Mdt$$
  $$165 = 6000 \times d \times \left(\dfrac{90}{360}\right)$$
  $$165 = 1500d$$
  $$\dfrac{165}{1500} = \dfrac{1500d}{1500}$$
  $$0.11 = d$$

The discount rate is 11%.

This is the simple interest rate used to calculate the simple interest due on the face value of the note.

Suppose that a supplier, upon delivery of merchandise to a customer, receives a 90-day promissory note from the customer in the amount of the invoice. The supplier, who would like to have the cash now in order to purchase goods, may go to a bank that will buy the promissory note and give the supplier cash. The bank purchases the note by first discounting the note. The supplier receives in cash the proceeds of a simple discount note.

Now suppose that the supplier does not go to the bank until 30 days after the 90-day note was signed. There are now 60 days left until payment is due on the note. The bank will calculate the discount on the days remaining. Therefore, in the simple discount formula $D = Mdt$, the discount is calculated using the number of days until the due date of the promissory note. This discount is subtracted from the face value of the note to determine the proceeds.

To summarize, when a promissory note is purchased with a simple discount note:

Find the interest and the maturity value of the promissory note.
Find the time until the maturity date of the promissory note.
Find the discount on the simple discount note.
    The maturity value M is the maturity value of the promissory note.
    The time t is the time until the maturity date of the note.
Find the proceeds (the maturity value minus the discount).

A real estate company owns a 5-year, 8% simple interest note that has a face value of $10,000. Two years before the note is due, the company sells the note to a bank by discounting it at 11%. How much does the bank pay for the note?

- Find the maturity value of
  the note. Recall that $I = Prt$.
  $P = 10,000$, $r = 8\% = 0.08$, $t = 5$

  $M = P + I$
  $= 10,000 + 10,000(0.08)(5)$
  $= 10,000 + 4000$
  $= 14,000$

- Use the simple discount formula
  to find the discount on a 2-year,
  11% note with a maturity value
  of 14,000.
  $M = 14,000$, $d = 11\% = 0.11$, $t = 2$

  $D = Mdt$
  $= 14,000(0.11)(2)$
  $= 3080$

- Find the proceeds of the $14,000
  note.
  $M = 14,000$, $D = 3080$

  $p = M - D$
  $= 14,000 - 3080$
  $= 10,920$

The proceeds are $10,920.

This means that the bank is paying $10,920 now to receive $14,000 in 2 years.

Example 1

The Pine Tree State Bank offers an electrical contractor a 120-day, 8% simple discount note that has a face value of $2400. Find the proceeds of the note.

Strategy

To find the proceeds:

- Use the formula $D = Mdt$ to find the discount.
  $M = 2400$, $d = 0.08$, $t = \dfrac{120}{360}$

- Use the formula $p = M - D$.

Solution

$D = Mdt$

$\quad = 2400(0.08)\left(\dfrac{120}{360}\right)$

$\quad = 64$

The discount is $64.

$p = M - D$
$\quad = 2400 - 64$
$\quad = 2336$

The proceeds are $2336.

You Try It 1

A swimming pool company signs a 45-day, 7.5% simple discount note that has a face value of $8000. Find the proceeds of the note.

Your strategy

Your solution

Solution on p. A34

Example 2

On June 18, the Morgan Company discounted at a bank a 6%, 120-day note for $5000 dated May 19. The bank's discount rate was 6.4%. Find the proceeds received by the Morgan Company.

Strategy

To find the proceeds:

- Use the formula $M = P + I$ to find the maturity value of the note. Recall $I = Prt$.

  $P = 5000, r = 0.06, t = \dfrac{120}{360}$

- Find the time until the maturity date of the note. (The Day of the Year Table is used in the solution that follows.)

- Use the formula $D = Mdt$ to find the discount on the simple discount note.
  M = the maturity value found in Step 1 of the solution, $d = 6.4\% = 0.064$, t = time in days (the number of days until the maturity date of the note, which was found in Step 2)

- Use the formula $p = M - D$ to find the proceeds.

Solution

$M = P + I$

$\quad = 5000 + 5000(0.06)\left(\dfrac{120}{360}\right)$

$\quad = 5000 + 100$

$\quad = 5100$

May 19 is Day 139.
June 18 is Day 169.

$169 - 139 = 30$ days from May 19 to June 18.
120 days − 30 days = 90 days

The maturity date of the note is in 90 days.

$D = Mdt$

$\quad = 5100(0.064)\left(\dfrac{90}{360}\right)$

$\quad = 81.60$

$p = M - D$
$\quad = 5100 - 81.60$
$\quad = 5018.40$

The proceeds received were $5018.40.

You Try It 2

On July 7, the Akron Company discounted at a bank an 8%, 90-day note for $7000 dated June 7. The bank's discount rate was 8.4%. Find the proceeds received by the Akron Company.

Your strategy

Your solution

Solution on p. A34

EXERCISE 10.2A

Solve.

1. A note that has a face value of $8000 and is due in 60 days is discounted 7.5%. Find the discount.

2. A note that has a face value of $2000 and is due in 45 days is discounted 9%. Find the discount.

3. A note that has a face value of $14,000 and is due in 9 months is discounted 7.5%. What is the discount?

4. A note that has a face value of $1800 and is due in 3 months is discounted 10%. What is the discount?

5. A simple discount note dated March 25 has a face value of $2000 and is due May 9. The discount rate is 7.5%. Find (a) the term of the note in days and (b) the discount.

6. A simple discount note dated July 18 has a face value of $5000 and is due August 17. The discount rate is 8%. Find (a) the term of the note in days and (b) the discount.

1. _____

2. _____

3. _____

4. _____

5. (a) _____

   (b) _____

6. (a) _____

   (b) _____

7. Find (a) the discount and (b) the proceeds of a $3300 simple discount note that is due in 2 months and has a discount rate of 12%.

8. Find (a) the discount and (b) the proceeds of a $4500 simple discount note that is due in 10 days and has a discount rate of 12%.

9. A simple discount note dated April 14 has a face value of $2500 and is due May 14. The discount rate is 9%. Find (a) the term of the note in months, (b) the discount, and (c) the proceeds.

10. A simple discount note dated February 22 has a face value of $11,000 and is due May 22. The discount rate is 8.4%. Find (a) the term of the note in months, (b) the discount, and (c) the proceeds.

11. A 60-day simple discount note that has a face value of $4500 is discounted $88. Find the discount rate. Round to the nearest tenth of a percent.

12. A 45-day simple discount note that has a face value of $5000 is discounted $50. Find the discount rate.

13. Heather Ross signs a simple discount note on December 12. The note has a face value of $7500 and is discounted $200. The maturity date of the note is April 12. Find the discount rate.

7. (a) _____

    (b) _____

8. (a) _____

    (b) _____

9. (a) _____

    (b) _____

    (c) _____

10. (a) _____

    (b) _____

    (c) _____

11. _____

12. _____

13. _____

14. Dellas King signs a simple discount note on August 8. The note has a face value of $6600 and is discounted $115.50. The maturity date of the note is November 8. Find the discount rate.

15. Kent Harding signs a simple discount note dated June 23. The note has a face value of $3000 and is due August 22. Find the proceeds if the discount rate is 8%.

16. Wayne Miller signs a simple discount note dated October 13. The note has a face value of $8000 and is due April 13. Find the proceeds if the discount rate is 8.5%.

17. A real estate firm owns a 5-year, 8% simple interest note that has a face value of $25,000. A bank agrees to purchase the note 2 years before it is due by discounting it at 10%. Find (a) the maturity value of the note, (b) the discount, and (c) the proceeds.

18. A building contractor sold an office building to a group of attorneys. As part of the purchase agreement, the contractor received a 2-year, 9% simple interest note that had a face value of $6000. Six months after receiving the note, the contractor sold the note to a bank after discounting it 10%. Find (a) the maturity value of the note, (b) the discount, and (c) the proceeds.

19. A real estate syndicate owns a 5-year, 10% simple interest note that has a face value of $5000. A bank agrees to purchase the note 3 years before it is due by discounting it at 8%. How much will the bank pay for the note?

14. _____

15. _____

16. _____

17. (a) _____

(b) _____

(c) _____

18. (a) _____

(b) _____

(c) _____

19. _____

20. Gary Deerfield owns a 3-year, 9% simple interest note that has a face value of $2500. Gary decides to sell the note after keeping it for 1 year. How much will the bank pay Gary for the note if the note is discounted 10%?

21. On January 26 the Abbott Company discounted at a bank an 8%, 120-day note for $18,000 dated November 27. The bank's discount rate was 8.8%. Find (a) the maturity value of the note, (b) the discount, and (c) the proceeds.

22. On August 29 the Belmont Company discounted at a bank a 9%, 90-day note for $12,000 dated June 30. The bank's discount rate was 10%. Find (a) the maturity value of the note, (b) the discount, and (c) the proceeds.

23. On May 3 a supplier discounted a $6000, 6%, 180-day promissory note at a bank. The promissory note was dated March 4. The discount rate was 6.6%. Find the proceeds received by the supplier.

24. On July 28 a retailer discounted a $4000, 5%, 60-day promissory note at a bank. The promissory note was dated June 28. The discount rate was 6%. Find the proceeds received by the retailer.

25. On September 18 the Tower Company accepted a $9000, 8%, 210-day promissory note. On November 3 the Tower Company discounted the note at a bank at a discount rate of 9%. What proceeds did the Tower Company receive?

20. _____

21. (a) _____

    (b) _____

    (c) _____

22. (a) _____

    (b) _____

    (c) _____

23. _____

24. _____

25. _____

SECTION 10.3   Compound Interest

Objective 10.3A   *To find the future value of compound interest*

If the interest received on the principal of an investment is added to the principal and the new amount is reinvested at the same interest rate, the interest is called **compound interest.** Compound interest is usually compounded annually (once a year), semiannually (twice a year), quarterly (four times a year), monthly (twelve times a year), or daily (once a day). The time period of each addition of interest to principal is the **conversion period** or **compounding period.**

An investor places $100 in an account that earns 8% annual interest compounded semiannually. Find the interest earned on the investment in 2 years.

♦ Calculate the interest earned each 6 months.

Year 1: First 6 months $I = Prt$

$P = 100, r = 0.08, t = \dfrac{6}{12} = \dfrac{1}{2}$ $= 100(0.08)\left(\dfrac{1}{2}\right) = 4$

New principal = $100 + $4
            = $104

Year 1: Second 6 months $I = Prt$

$P = 104, r = 0.08, t = \dfrac{1}{2}$ $= 104(0.08)\left(\dfrac{1}{2}\right) = 4.16$

New principal = $104 + $4.16
            = $108.16

Year 2: First 6 months $I = Prt$

$P = 108.16, r = 0.08, t = \dfrac{1}{2}$ $= 108.16(0.08)\left(\dfrac{1}{2}\right) \approx 4.33$

New principal = $108.16 + $4.33
            = $112.49

Year 2: Second 6 months $I = Prt$

$P = 112.49, r = 0.08, t = \dfrac{1}{2}$ $= 112.49(0.08)\left(\dfrac{1}{2}\right) \approx 4.50$

New principal = $112.49 + $4.50
            = $116.99

♦ Subtract the original value of the $116.99 - 100 = 16.99$
investment ($100) from the new
value ($116.99).

The interest earned is $16.99.

The effects of compounding interest $I = Prt$
can be illustrated by computing the $= 100(0.08)(2) = 16$
simple interest on the same $100 in-
vested for 2 years at an annual simple
interest rate of 8%.

The simple interest earned is $16, which is $.99 less than the compound interest.

TABLE 10.1   Compound Interest Table

$i\% = \dfrac{\text{annual interest rate}}{\text{compounding periods per year}}$, $n = \text{compounding periods per year} \times \text{number of years}$

n	0.5%	$\frac{2}{3}$%	0.75%	1%	1.5%	2%	2.5%	3%	4%
1	1.005000	1.006667	1.007500	1.010000	1.015000	1.020000	1.025000	1.030000	1.040000
2	1.010025	1.013378	1.015056	1.020100	1.030225	1.040400	1.050625	1.060900	1.081600
3	1.015075	1.020134	1.022669	1.030301	1.045678	1.061208	1.076891	1.092727	1.124864
4	1.020151	1.026935	1.030339	1.040604	1.061364	1.082432	1.103813	1.125509	1.169859
5	1.025251	1.033781	1.038067	1.051010	1.077284	1.104081	1.131408	1.159274	1.216653
6	1.030378	1.040673	1.045852	1.061520	1.093443	1.126162	1.159693	1.194052	1.265319
7	1.035529	1.047610	1.053696	1.072135	1.109845	1.148686	1.188686	1.229874	1.315932
8	1.040707	1.054595	1.061599	1.082857	1.126493	1.171659	1.218403	1.266770	1.368569
9	1.045911	1.061625	1.069561	1.093685	1.143390	1.195093	1.248863	1.304773	1.423312
10	1.051140	1.068703	1.077583	1.104622	1.160541	1.218994	1.280085	1.343916	1.480244
11	1.056396	1.075827	1.085664	1.115668	1.177949	1.243374	1.312087	1.384234	1.539454
12	1.061678	1.083000	1.093807	1.126825	1.195618	1.268242	1.344889	1.425761	1.601032
13	1.066986	1.090220	1.102010	1.138093	1.213552	1.293607	1.378511	1.468534	1.665074
14	1.072321	1.097488	1.110276	1.149474	1.231756	1.319479	1.412974	1.512590	1.731676
15	1.077683	1.104804	1.118603	1.160969	1.250232	1.345868	1.448298	1.557967	1.800944
16	1.083071	1.112170	1.126992	1.172579	1.268986	1.372786	1.484506	1.604706	1.872981
17	1.088487	1.119584	1.135445	1.184304	1.288020	1.400241	1.521618	1.652848	1.947900
18	1.093929	1.127048	1.143960	1.196147	1.307341	1.428246	1.559659	1.702433	2.025817
19	1.099399	1.134562	1.152540	1.208109	1.326951	1.456811	1.598650	1.753506	2.106849
20	1.104896	1.142125	1.161184	1.220190	1.346855	1.485947	1.638616	1.806111	2.191123
21	1.110420	1.149740	1.169893	1.232392	1.367058	1.515666	1.679582	1.860295	2.278768
22	1.115972	1.157404	1.178667	1.244716	1.387564	1.545980	1.721571	1.916103	2.369919
23	1.121552	1.165120	1.187507	1.257163	1.408377	1.576899	1.764611	1.973587	2.464716
24	1.127160	1.172888	1.196414	1.269735	1.429503	1.608437	1.808726	2.032794	2.563304
36	1.196681	1.270237	1.308645	1.430769	1.709140	2.039887	2.432535	2.898278	4.103933
48	1.270489	1.375666	1.431405	1.612226	2.043478	2.587070	3.271490	4.132252	6.570528
60	1.348850	1.489846	1.565681	1.816697	2.443220	3.281031	4.399790	5.891603	10.519627
120	1.819397	2.219640	2.451357	3.300387	5.969323	10.765163	19.358150	34.710987	110.662561

n	5%	6%	7%	8%	9%	10%	11%	12%	13%
1	1.050000	1.060000	1.070000	1.080000	1.090000	1.100000	1.110000	1.120000	1.130000
2	1.102500	1.123600	1.144900	1.166400	1.188100	1.210000	1.232100	1.254400	1.276900
3	1.157625	1.191016	1.225043	1.259712	1.295029	1.331000	1.367631	1.404928	1.442897
4	1.215506	1.262477	1.310796	1.360489	1.411582	1.464100	1.518070	1.573519	1.630474
5	1.276282	1.338226	1.402552	1.469328	1.538624	1.610510	1.685058	1.762342	1.842435
6	1.340096	1.418519	1.500730	1.586874	1.677100	1.771561	1.870415	1.973823	2.081952
7	1.407100	1.503630	1.605781	1.713824	1.828039	1.948717	2.076160	2.210681	2.352605
8	1.477455	1.593848	1.718186	1.850930	1.992563	2.143589	2.304538	2.475963	2.658444
9	1.551328	1.689479	1.838459	1.999005	2.171893	2.357948	2.558037	2.773079	3.004042
10	1.628895	1.790848	1.967151	2.158925	2.367364	2.593742	2.839421	3.105848	3.394567
11	1.710339	1.898299	2.104852	2.331639	2.580426	2.853117	3.151757	3.478550	3.835861
12	1.795856	2.012196	2.252192	2.518170	2.812665	3.138428	3.498451	3.895976	4.334523
13	1.885649	2.132928	2.409845	2.719624	3.065805	3.452271	3.883280	4.363493	4.898011
14	1.979932	2.260904	2.578534	2.937194	3.341727	3.797498	4.310441	4.887112	5.534753
15	2.078928	2.396558	2.759032	3.172169	3.642482	4.177248	4.784589	5.473566	6.254270
16	2.182875	2.540352	2.952164	3.425943	3.970306	4.594973	5.310894	6.130394	7.067326
17	2.292018	2.692773	3.158815	3.700018	4.327633	5.054470	5.895093	6.866041	7.986078
18	2.406619	2.854339	3.379932	3.996019	4.717120	5.559917	6.543553	7.689966	9.024268
19	2.526950	3.025600	3.616528	4.315701	5.141661	6.115909	7.263344	8.612762	10.197423
20	2.653298	3.207135	3.869684	4.660957	5.604411	6.727500	8.062312	9.646293	11.523088
21	2.785963	3.399564	4.140562	5.033834	6.108808	7.400250	8.949166	10.803848	13.021089
22	2.925261	3.603537	4.430402	5.436540	6.658600	8.140275	9.933574	12.100310	14.713831
23	3.071524	3.819750	4.740530	5.871464	7.257874	8.954302	11.026267	13.552347	16.626629
24	3.225100	4.048935	5.072367	6.341181	7.911083	9.849733	12.239157	15.178629	18.788091
36	5.791816	8.147252	11.423942	15.968172	22.251225	30.912681	42.818085	59.135574	81.437412
48	10.401270	16.393872	25.728907	40.210573	62.585237	97.017234	149.796954	230.390776	352.992345
60	18.679186	32.987691	57.946427	101.257064	176.031292	304.481640	524.057242	897.596933	1530.053473

TABLE 10.2   Daily Compounding Table

n	6%	7%	8%	9%	10%	11%	12%	13%	14%
1	1.061831	1.072501	1.083277	1.094162	1.105156	1.116259	1.127474	1.138802	1.140000
2	1.127486	1.150258	1.173490	1.197190	1.221369	1.246035	1.271198	1.296869	1.299600
3	1.197199	1.233653	1.271215	1.309920	1.349803	1.390898	1.433243	1.476877	1.481544
4	1.271224	1.323094	1.377079	1.433265	1.491742	1.552603	1.615945	1.681870	1.688960
5	1.349825	1.419019	1.491758	1.568224	1.648607	1.733107	1.821937	1.915316	1.925415
6	1.433286	1.521899	1.615988	1.715891	1.821967	1.934597	2.054187	2.181165	2.194973
7	1.521908	1.632238	1.750564	1.877463	2.013557	2.159512	2.316043	2.483914	2.502269
8	1.616010	1.750577	1.896346	2.054248	2.225294	2.410576	2.611279	2.828686	2.852586
9	1.715930	1.877496	2.054269	2.247680	2.459296	2.690828	2.944150	3.221312	3.251949
10	1.822028	2.013616	2.225343	2.459326	2.717904	3.003661	3.319453	3.668436	3.707221

The method we just used to compute compound interest would be very tedious if the number of compounding periods were large. Fortunately, tables and formulas can be used for these calculations. We will first discuss using a compound interest table and then introduce the formula for finding compound interest using a calculator.

When using a compound interest table to calculate the value of an investment earning compound interest, use the following formula:

Compound Interest Formula for Use with Tables

$$FV = PV \times \text{compounding factor}$$

FV = future value
PV = present value

The **future value, FV,** of an investment is the maturity value of the investment after the original principal has been invested for a given period of time. The **present value, PV,** is the original principal invested.

Table 10.1 gives the **compounding factor** for various interest rates and time periods. To use this table, find $i\%$, the interest rate per compounding period, and n, the total number of compounding periods.

$$i\% = \frac{r}{c} = \frac{\text{annual interest rate}}{\text{compounding periods per year}}$$

$$n = ct = \text{compounding periods per year} \times \text{number of years}$$

The compounding factor is the number found where the row for n and the column for $i\%$ intersect. The compounding factor for $i\% = 0.75\%$ and $n = 48$ (1.431405) is highlighted in the table.

To illustrate determining the values of i and n, consider an account earning 8% interest, compounded quarterly, for a period of 3 years.

The annual interest rate is 8%: $r = 8\%$
When interest is compounded quarterly,
there are 4 compounding periods per year: $c = 4$
The time is 3 years. $t = 3$

$$i\% = \frac{r}{c} = \frac{8\%}{4} = 2\%$$
$$n = ct = 4(3) = 12$$

It may be helpful to summarize the possible values of c.

If interest is	c is
compounded annually	1
compounded semiannually	2
compounded quarterly	4
compounded monthly	12
compounded daily	360

Note that the Compound Interest Table gives i as a percent. Therefore, when using the table to calculate the future value of an investment, leave the annual interest rate as a percent; do not rewrite it as a decimal. Also, do not round a number given in the table; perform calculations with all of the given digits.

An investor places $100 in an account that earns 8% annual interest compounded semiannually. Find the future value of the investment in 2 years. (Note that this is the same problem that we examined on page 299.)

◆ Use Table 10.1 to find the compounding factor.
PV = 100, $n = ct = 2(2) = 4$,

$$i\% = \frac{r}{c} = \frac{8\%}{2} = 4\%$$

FV = PV × compounding factor
= 100 × 1.169859
= 116.9859

The compounding factor for $n = 4$ and $i = 4$ is 1.169859.

The future value of the investment is $116.99.

The Compound Interest Table on page 300 does not include all possible values of i or n. The compound interest formula that follows can be used to calculate the future value of an investment at any given interest rate for any number of compounding periods per year.

Compound Interest Formula

$$\mathbf{FV = PV(1 + i)^n}$$

FV = future value of an investment
PV = present value of an investment

$$i = \frac{r}{c} = \frac{\text{annual interest rate}}{\text{compounding periods per year}}$$

$$n = ct = \text{compounding periods per year} \times \text{number of years}$$

The calculator keystrokes involved in calculating the future value of an investment using this compound interest formula are given on page 313.

When the Compound Interest Table is used to calculate future value, the interest rate is left as a percent. When the compound interest formula and a calculator are used to calculate future value, the interest rate is written as a decimal.

Find the future value of $10,000 deposited in an account earning 8% interest compounded semiannually for 5 years.

◆ Use the compound interest formula.

PV = 10,000, $n = ct = 2(5) = 10$,

$$i = \frac{r}{c} = \frac{8\%}{2} = \frac{0.08}{2} = 0.04$$

$FV = PV(1 + i)^n$
$= 10,000(1 + 0.04)^{10}$

$= 10,000(1.4802443)$
$\approx 14,802.44$

The future value of the investment is $14,802.44.

In the Compound Interest Table, the compounding factor when $i = 4\%$ and $n = 10$ is 1.480244, which is very close to the number 1.4802443 we found by using a calculator. This is because the Compound Interest Table gives the value of $(1 + i)^n$ for different values of i and n. The Compound Interest Table was created using the compound interest formula.

The value of $(1 + i)^n$ is the future value of $1. It is multiplied by the number of dollars invested, the present value, to determine the future value of the investment.

Calculate the future value of $5000 earning 9% interest compounded daily for 3 years by (a) using the Compound Interest Table and (b) using a calculator.

(a) Use Table 10.2 on page 300
to find the compounding
factor for 3 years at 9%.
PV = 5000

FV = PV × compounding factor
= 5000 × 1.309920
= 6549.60

The future value of the investment is $6549.60.

(b) Use the compound interest
formula.
PV = 5000,
$n = ct = 360(3) = 1080$,
$i = \dfrac{r}{c} = \dfrac{0.09}{360} = 0.00025$

$FV = PV(1 + i)^n$
$= 5000(1 + 0.00025)^{1080}$
$= 5000(1.3099203)$
≈ 6549.60

The future value of the investment is $6549.60.

To find the interest earned on an investment earning compound interest, subtract the present value of the investment from the future value.

Compound Interest Earned

CI = FV − PV

CI = compound interest
FV = future value of an investment
PV = present value of an investment

For the investment in the previous example,
the compound interest earned is $1549.60.

CI = FV − PV
= 6549.60 − 5000
= 1549.60

Find the interest earned in 3 years on an investment of $2000 at an annual interest rate of 6% compounded monthly. Use the Compound Interest Table.

◆ Find the future value.
PV = 2000, $n = ct = 12(3) = 36$

$i\% = \dfrac{r}{c} = \dfrac{6\%}{12} = 0.5\%$

FV = PV × compounding factor
= 2000 × 1.196681
≈ 2393.36

◆ Find the interest earned.

CI = FV − PV
= 2393.36 − 2000
= 393.36

The interest earned is $393.36.

Either the compound interest formula and a calculator or the Compound Interest Table can be used to calculate the future value of an investment. The answer will be the same. (Occasionally the answers will differ by a cent or by a few cents, because the compounding factors in the table are rounded values.) In the examples that follow, the Compound Interest Table is used. If you are using the compound interest formula and a calculator, the compounding factor shown in the solution is the value of $(1 + i)^n$ in the formula.

Example 1

An investment of $10,000 is placed in an account that earns 9% annual interest compounded monthly. Find the future value of the investment in 10 years.

Strategy

To find the future value, use the formula FV = PV × compounding factor. Use Table 10.1 to find the compounding factor.

$$PV = 10{,}000, \; i\% = \frac{r}{c} = \frac{9\%}{12} = 0.75\%,$$

$$n = ct = 12(10) = 120$$

Solution

$$
\begin{aligned}
FV &= PV \times \text{compounding factor} \\
&= 10{,}000 \times 2.451357 \\
&= 24{,}513.57
\end{aligned}
$$

The future value of the investment is $24,513.57.

You Try It 1

A self-employed individual places $5000 in a Keogh account that earns 8% annual interest compounded monthly. Find the future value of the investment in 10 years.

Your strategy

Your solution

Example 2

Rachel Mazur invests $5000 in an account that earns 8% annual interest compounded daily. How much interest will Rachel's investment earn in 10 years?

Strategy

To find the interest earned in 10 years:

◆ Find the future value of the investment. Use the formula FV = PV × compounding factor. Use Table 10.2 to find the compounding factor. PV = 5000, 8% daily compounding for 10 years

◆ Use the formula CI = FV − PV.

Solution

$$
\begin{aligned}
FV &= PV \times \text{compounding factor} \\
&= 5000 \times 2.225343 \\
&\approx 11{,}126.72
\end{aligned}
$$

$$
\begin{aligned}
CI &= FV - PV \\
&= 11{,}126.72 - 5000 \\
&= 6126.72
\end{aligned}
$$

The interest earned will be $6126.72.

You Try It 2

Spencer Liston invests $12,000 in an account that earns 9% annual interest compounded daily. How much interest will Spencer's investment earn in 5 years?

Your strategy

Your solution

Solutions on p. A35

Objective 10.3B   *To find the present value of compound interest*

Recall that future value is the maturity value of an investment after the original principal has been invested for a certain period of time. The present value is the original principal invested, or the value of the investment now before it earns any interest. Present value is used to find how much money must be invested today in order for the investment to have a specific value at a future date.

The present value of an investment can be found either by using a calculator or by using the Compound Interest Table. When using the Compound Interest Table, use the following formula:

> **Present Value Formula for Use with Tables**
>
> **PV = FV ÷ compounding factor**   PV = present value
>   FV = future value

The Sampson Company wishes to have $30,000 in 5 years in order to purchase new equipment. In order to reach this goal, how much money must the company deposit now in an account that earns 8% annual interest compounded quarterly?

◆ We want to find the present value
   of $30,000. Use Table 10.1 to
   find the compounding factor.
   FV = 30,000, $n = ct = 4(5) = 20$,
   $i\% = \dfrac{r}{c} = \dfrac{8\%}{4} = 2\%$

   PV = FV ÷ compounding factor
   $= 30,000 \div 1.485947$
   $\approx 20,189.15$

The present value of the investment is $20,189.15.

The company should deposit $20,189.15 now in order to have $30,000 in 5 years.

The present value formula that follows can be used to calculate the future value of an investment at any given interest rate for any number of compounding periods per year.

> **Present Value Formula**
>
> $$PV = \dfrac{FV}{(1 + i)^n}$$
>
> PV = present value of an investment
> FV = future value of an investment
> $i = \dfrac{r}{c} = \dfrac{\text{annual interest rate}}{\text{compounding periods of year}}$
>
> $n = ct$ = compounding periods per year × number of years

The calculator keystrokes involved in calculating the present value of an investment by using the present value formula are given on page 313.

Remember that when the Compound Interest Table is used to calculate present value, the interest rate is left as a percent. When the present value formula and a calculator are used to calculate present value, the interest rate is written as a decimal.

How much money should be invested in an account that earns 9% annual interest compounded semiannually in order to have $20,000 in 5 years? Use the present value formula.

$FV = 20,000$, $n = ct = 2(5) = 10$,

$i = \dfrac{r}{c} = \dfrac{0.09}{2} = 0.045$

$$PV = \frac{FV}{(1 + i)^n}$$

$$= \frac{20,000}{(1 + 0.045)^{10}}$$

$$= \frac{20,000}{1.5529694}$$

$$\approx 12,878.55$$

$12,878.55 should be invested in the account now in order to have $20,000 in 5 years.

The Ashland Company will need to have $10,000 in 2 years in order to replace machinery. Determine how much money must be deposited in an account that earns 9% annual interest compounded daily by (a) using the Compound Interest Table and (b) using the present value formula.

(a) Use Table 10.2 on page 300 to find the compounding factor for 2 years at 9%.
$FV = 10,000$

$PV = FV \div$ compounding factor
$= 10,000 \div 1.197190$
≈ 8352.89

$8352.89 should be deposited now.

(b) Use the present value formula.
$FV = 10,000$,

$n = ct = 360(2) = 720$,

$i = \dfrac{r}{c} = \dfrac{0.09}{360} = 0.00025$

$$PV = \frac{FV}{(1 + i)^n}$$

$$= \frac{10,000}{(1 + 0.00025)^{720}}$$

$$= \frac{10,000}{1.1971904}$$

$$\approx 8352.89$$

$8352.89 should be deposited now.

Either the present value formula or the Compound Interest Table can be used to calculate the present value of an investment. As in the calculation of future value, the answers occasionally differ by up to a few cents because the compounding factors in the table are rounded values.

In the examples that follow, the Compound Interest Table is used. If you are using the present value formula, the value of $(1 + i)^n$ is the compounding factor shown in the solution.

Example 3

How much money should be invested in an account that earns 9% annual interest compounded monthly in order to have $7000 in 5 years?

Strategy

To find the amount that should be invested, use the formula PV = FV ÷ compounding factor. Use Table 10.1 to find the compounding factor.
FV = 7000, $n = ct = 12(5) = 60$,
$i\% = \dfrac{r}{c} = \dfrac{9\%}{12} = 0.75\%$

Solution

PV = FV ÷ compounding factor
    = 7000 ÷ 1.565681
    = 4470.898

$4470.90 should be invested.

You Try It 3

How much money should be invested in an account that earns 6% annual interest compounded semiannually in order to have $20,000 in 4 years?

Your strategy

Your solution

Solution on p. A35

Objective 10.3C   *To calculate effective interest rates*

When interest is compounded, the annual rate of interest is called the **nominal rate.** The **effective rate** is the simple interest rate that would yield the same amount of interest after one year. When a bank advertises a "7% annual interest rate compounded daily and yielding 7.25%," the nominal interest rate is 7% and the effective rate is 7.25%.

Consider $100 deposited at 6% compounded monthly for one year.

The future value after one year is $106.17.

FV = PV × compounding factor
   = 100 × 1.061678
   ≈ 106.17

The interest earned is $6.17.

CI = FV − PV
   = 106.17 − 100
   = 6.17

Now consider $100 deposited at an annual simple interest rate of 6.17%.

The interest earned in one year is $6.17.

$I = Prt$
   = 100(0.0617)(1)
   = 6.17

The interest earned on $100 is the same when it is deposited at 6% compounded monthly and when it is deposited at an annual simple interest rate of 6.17%. 6.17% is the effective annual rate of 6% compounded monthly.

In the previous example, \$100 was used as the principal. Using \$100 for *P* results in multiplying the interest rate by 100. Remember that the interest rate is written as a decimal in the equation *I = Prt*, and a decimal is written as a percent by multiplying by 100. Therefore, when *P* = 100, the interest earned on the investment (\$6.17) is the same number as the effective annual rate (6.17%).

A savings and loan association offers a certificate of deposit at an annual interest rate of 8% compounded daily. Find the effective rate to the nearest hundredth of a percent.

- Use Table 10.2 on page 300 to find the future value of \$100.
  PV = 100, 8% daily compounding for 1 year

  The future value of \$1 is \$108.3277.

$$FV = PV \times \text{compounding factor}$$
$$= 100 \times 1.083277$$
$$= 108.3277$$

- Find the interest earned.
  FV = 108.3277, PV = 100

$$CI = FV - PV$$
$$= 108.3277 - 100$$
$$= 8.3277$$

The effective rate is 8.33%.

To compare two investments or loan agreements, calculate the effective annual rate of each. When you are considering investing your money, you want the highest effective rate so that your money will earn more interest. If you are considering borrowing money, you want the lowest effective rate so that you will pay less interest on the loan.

Example 4

A bank offers a certificate of deposit at an annual interest rate of 8% compounded monthly. Find the effective rate to the nearest hundredth of a percent.

Strategy

To find the effective rate:

- Find the future value of \$100.
  PV = 100, *n = ct* = 12(1) = 12,
  $i\% = \dfrac{r}{c} = \dfrac{8\%}{12} = \dfrac{2}{3}\%$

- Find the interest earned on the \$100.

Solution

$$FV = PV \times \text{compounding factor}$$
$$= 100 \times 1.083000$$
$$= 108.3000$$

$$CI = FV - PV$$
$$= 108.3000 - 100$$
$$= 8.3000$$

The effective rate is 8.30%.

You Try It 4

A credit union offers a certificate of deposit at an annual interest rate of 7% compounded daily. Find the effective rate to the nearest hundredth of a percent.

Your strategy

Your solution

Solution on p. A35

EXERCISE 10.3A

Solve.

1. An architect deposited $10,000 in an account that earns 8% interest compounded daily. Find the future value of the investment in 2 years.

2. A firefighter invested $3500 in an account that earns 12% interest compounded monthly. Find the future value of the investment in 3 years.

3. An accountant invested $2300 in an account that earns 10% interest compounded quarterly. Find the future value of the investment in 6 years.

4. A mechanic invested $3200 in an account that earns 10% interest compounded daily. Find the future value of the investment in 5 years.

5. A nurse invested $4500 in an account that earns 8% compounded daily. Find the future value of the investment in 4 years.

6. A corporation predicts that its earnings per share will increase at an annual compounding rate of 10%. The corporation's earnings per share this year are $4.35. Find the future value of the earnings in 5 years.

7. An analyst predicts that the earnings per share of a corporation will increase at an annual compounding rate of 8%. The corporation's earnings per share this year are $6.24. Find the future value of the earnings in 3 years.

8. Shirley Clements deposits $3200 in an account that earns 9% annual interest compounded monthly. Find the interest earned on the account in 4 years.

9. Tie Masotta deposits $8500 in an account that earns 9% annual interest compounded daily. Find the interest earned in 5 years.

1. _____

2. _____

3. _____

4. _____

5. _____

6. _____

7. _____

8. _____

9. _____

10. A medical lab technician deposited $5500 in an account that earns 7% annual interest compounded daily. Find the interest earned in 4 years.

An investment that has a present value of $1000 is placed in an account that earns 8% annual interest for 1 year. Complete the following table for the compounding periods.

Compounding Period	Compounding Factor	Present Value	Future Value
Annually	11.	$1000	12.
Semiannually	13.	$1000	14.
Quarterly	15.	$1000	16.
Monthly	17.	$1000	18.
Daily	19.	$1000	20.

Solve.

21. Use the table above to find the difference between the amount of interest earned in 1 year from daily compounding of interest and the amount of interest earned from annual compounding.

22. Use the table above to find the difference between the amount of interest earned from daily compounding and the amount of interest earned from monthly compounding.

EXERCISE 10.3B

Solve.

23. How much money should Tina Cortez invest in an account that earns 9% annual interest compounded daily in order to have $2000 in 5 years?

24. How much money should Arthur Smith invest in an account that earns 6% annual interest compounded quarterly in order to have $5000 in 4 years?

25. A company that produces computer components wishes to have $15,000 4 years from now in order to purchase new office equipment. How much money must it deposit now in an account that earns 8% annual interest compounded daily in order to reach this goal?

26. A company that produces computer software wishes to have $35,000 in 2 years. How much should it deposit into an account that earns 8% compounded quarterly to reach that goal?

10. _____
11. _____
12. _____
13. _____
14. _____
15. _____
16. _____
17. _____
18. _____
19. _____
20. _____
21. _____
22. _____
23. _____
24. _____
25. _____
26. _____

27. A federal government investment will have a future value of $5000 in 5 years. What is the present value of the investment if the annual interest rate is 8% compounded semiannually?

27. _____

28. _____

29. _____

30. _____

31. _____

32. _____

33. _____

28. A state government offers an investment that will have a future value of $10,000 in 10 years. What is the present value of the investment if the annual interest rate is 6% compounded semiannually?

29. A savings and loan association offers an investment that will have a future value of $5000 in 5 years. What is the present value of the investment if the annual interest rate is 7% compounded daily?

30. A bank offers an investment that will have a future value of $2500 in 3 years. What is the present value of the investment if the annual interest rate is 9% compounded monthly?

31. An investor must choose between two investments. One investment has a future value of $10,000 in 8 years. The second investment has a future value of $7500 in 6 years. Which investment has the greater present value if the annual interest rate is 7% compounded daily?

32. An investor must choose between two investments. One investment has a future value of $5000 in 6 years. The second investment has a future value of $9000 in 10 years. Which investment has the greater present value if the annual interest rate is 8% compounded daily?

33. A savings and loan association offers its depositors two different certificates of deposit. The first certificate has a future value of $5000 in 5 years. The second certificate has a future value of $7500 in 8 years. Which certificate of deposit has the greater present value if the annual interest rate is 9% compounded daily?

EXERCISE 10.3C

Solve.

34. A credit union offers a certificate of deposit at an annual interest rate of 7% compounded daily. Find the effective rate of interest.

35. A bank offers a certificate of deposit at an annual rate of 8% compounded quarterly. Find the effective rate of interest.

36. Beth Chipman has money in a savings account that earns an annual interest rate of 8% compounded monthly. What is the effective rate of interest on Beth's account?

37. Blake Hamilton has money in a savings account that earns an annual interest rate of 6% compounded monthly. What is the effective rate of interest on Blake's account?

Complete the following table for an annual interest rate of 8% for 1 year.

Compounding Period	Compounding Factor	Present Value	Future Value	Effective Rate
Annually	38.	$1.00	39.	40.
Semiannually	41.	$1.00	42.	43.
Quarterly	44.	$1.00	45.	46.
Monthly	47.	$1.00	48.	49.
Daily	50.	$1.00	51.	52.

53. Which has the higher effective annual rate, 6% compounded quarterly or 6.25% compounded semiannually? (Use a calculator to find the future value.)

54. Which has the higher effective annual rate, 7.8% compounded monthly or 7.5% compounded daily? (Use a calculator to find the future value.)

34. _____

35. _____

36. _____

37. _____

38. _____

39. _____

40. _____

41. _____

42. _____

43. _____

44. _____

45. _____

46. _____

47. _____

48. _____

49. _____

50. _____

51. _____

52. _____

53. _____

54. _____

CALCULATORS

The y^x Key and the $\frac{1}{x}$ Key

The y^x key is used to calculate a power of a number. For example, to evaluate 8^7, enter 8, press y^x, and enter 7. The display should read 2097152.

The y^x key is used in solving compound interest problems. Mastering the use of this key will enable you to calculate the present value and future value for *any* compound interest rate.

The **Compound Interest Formula** is

$$FV = PV(1 + i)^n$$

where $i = \dfrac{r}{c} = \dfrac{\text{annual interest rate}}{\text{compounding periods per year}}$

$n = ct = $ compounding periods per year $\times$ number of years

A deposit of $500 is made in an account that earns 6.5% annual interest compounded daily. Find the future value of the account in 5 years.

$PV = 500$, $i = 0.065 \div 360$, $n = 360(5) = 1800$

Enter	**Comments**
0.065 $\div$ 360 $=$	Calculate i.
$+$ 1 $=$	Calculate $1 + i$.
y^x 1800 $=$	Calculate $(1 + i)^n$.
$\times$ 500 $=$	Multiply by the deposit. Round to the nearest hundredth.

In 5 years, $692.00 will be in the account.

Note from the previous example that the keystroking sequence is

r $\div$ c $=$ $+$ 1 $=$ y^x n $=$ $\times$ PV $=$

Calculate the future value of $4000 deposited in an account earning 6% interest compounded monthly for 2 years.

Use the following keystrokes:

0.06 $\div$ 12 $=$ $+$ 1 $=$ y^x 24 $=$ $\times$ 4000 $=$

The future value of the investment is $4508.64.

The **Present Value Formula** is

$$PV = \frac{FV}{(1 + i)^n}$$

Note that the denominator is the same as the multiplier of PV in the compound interest formula. Calculate this expression as you would for the compound interest formula. Then use the $\frac{1}{x}$ key. The $\frac{1}{x}$ key takes the reciprocal of the number in the display. In effect, it places the number in the display in the denominator.

How much must be deposited in an account that earns 7.2% annual interest compounded daily in order to have $5000 in 4 years?

$$FV = 5000, \quad i = 0.072 \div 360, \quad n = 360(4) = 1440$$

Enter	**Comments**
0.072 $\boxed{\div}$ 360 $\boxed{=}$	Calculate i.
$\boxed{+}$ 1 $\boxed{=}$	Calculate $1 + i$.
$\boxed{y^x}$ 1440 $\boxed{=}$	Calculate $(1 + i)^n$.
$\boxed{\frac{1}{x}}$	Put the result in the denominator.
$\boxed{\times}$ 5000 $\boxed{=}$	Multiply by the future value. Round to the nearest hundredth.

A deposit of $3748.92 will have a future value of $5000 in 4 years.

Note from the previous example that the keystroking sequence is

$$r \boxed{\div} c \boxed{=} \boxed{+} 1 \boxed{=} \boxed{y^x} n \boxed{=} \boxed{\frac{1}{x}} \boxed{\times} FV \boxed{=}$$

How much money should be deposited in an account that earns 9% compounded semiannually in order to have $18,000 in 15 years?

Use the following keystrokes:

$$0.09 \boxed{\div} 2 \boxed{=} \boxed{+} 1 \boxed{=} \boxed{y^x} 30 \boxed{=} \boxed{\frac{1}{x}} \boxed{\times} 18000 \boxed{=}$$

$4806.00 should be deposited in the account.

BUSINESS CASE STUDY

Taking Advantage of Cash Discounts

You own and operate a retail appliance store. The enterprise is doing well, but profits have not been increasing over the past few years as much as you had anticipated they would. You are eager to expand your product line, but this will require extra cash and you are uncertain whether the business can handle the expansion at this time. You decide to hire an accountant to look over your books and make recommendations.

After examining your books, the accountant meets with you to discuss the findings. The first thing the accountant discusses with you is the practice of taking advantage of cash discounts.

During the past year alone, the accountant informs you, $450,000 worth of appliances have been ordered from wholesalers. On $300,000 of these orders, a 2% cash discount applied. On $150,000 of the orders, a 3% cash discount applied. Yet all of these bills were paid just before the final due date, which is usually 30 days after the receipt of the bill. "Two or three percent may not sound like a lot," the accountant says, "but it can make quite a difference in the amount you pay over a period of a year."

In order to start taking advantage of cash discounts now, you would, in effect, need to pay two months' worth of bills this month—the bills due after 30 days (which are usually paid at this time) *and* the more recent bills that must be paid promptly in order to receive the cash discount.

The accountant informs you that your monthly orders for appliances are fairly steady. Consequently, each monthly order during the past year was approximately $37,500 ($450,000 ÷ 12). Paying two months' bills this month would therefore require taking approximately $37,500 out of the business's cash investments. The money the accountant recommends be used for this purpose has been invested this past year at a 9% annual interest rate compounded monthly.

1. How much money would not have been spent on merchandise this past year if the company had taken advantage of the cash discounts available?

2. How much money would the business have lost by not earning interest on the $37,500 during the past year?

3. Find the difference between the amount not spent (the answer to Question 1) and the amount earned in interest (the answer to Question 2).

4. On the basis of the information provided, do you think you should take the accountant's advice and begin taking advantage of cash discounts?

CHAPTER SUMMARY

Key Words The **principal** is the amount of money deposited or borrowed. **Interest** is the amount paid for the use of borrowed money. The **interest rate** is the percent used to determine the amount of interest to be paid. Interest computed on the original principal is **simple interest.** (Objective 10.1A)

A **promissory note,** or simply a **note,** is a contract signed by a borrower for a loan. The principal of a note is the **face value** of the note. The **maturity value** is the sum of the principal and the interest. The **maturity date,** or **due date,** is the date on which the loan and interest must be completely repaid. The length of time between the date a note is written and its due date is the **term** of the note. (Objective 10.1B)

A **simple discount note** is a note in which the interest due is subtracted from the face value of the note. The borrower receives the face value minus the interest; this amount is the **proceeds** of the note. For a simple discount note, the face value and the maturity value are the same; both equal the proceeds plus the interest. The interest due on the face value of a simple discount note is called the **discount.** The interest rate used to calculate the interest due is the **discount rate.** (Objective 10.2A)

Compound interest is the interest computed on interest earned as well as on the original principal. The **conversion period,** or **compounding period,** is the time period of each addition of interest to the principal. (Objective 10.3A)

The **future value (FV)** of an investment is the maturity value of the investment after the original principal has been invested for a given period of time. The **present value (PV)** is the original principal invested. (Objective 10.3A)

When interest is compounded, the annual rate of interest is called the **nominal rate.** The **effective rate** is the simple interest rate that would yield the same amount of interest as the nominal rate after one year. (Objective 10.3C)

Essential Rules

Simple Interest Formula

$I = Prt$,
where I is the simple interest, P is the principal, r is the annual interest rate, and t is the time of the loan in years (Objective 10.1A)

To convert time to years:

Time in months: $t = \dfrac{\text{number of months}}{12}$

Time in days (exact method): $t = \dfrac{\text{number of days}}{365}$

Time in days (ordinary method): $t = \dfrac{\text{number of days}}{360}$ (Objective 10.1A)

Simple Discount Notes

$D = Mdt$ and $p = M - D$,
where D is the discount, M is the maturity value or face value, d is the discount rate, t is the time in years, and p is the proceeds (Objective 10.2A)

Compound Interest Formula for Use with Tables

$FV = PV \times$ compounding factor (Objective 10.3A)

Compound Interest Formula

$FV = PV(1 + i)^n$,

where $i = \dfrac{r}{c} = \dfrac{\text{annual interest rate}}{\text{compounding periods per year}}$ and

$n = ct =$ compounding periods per year $\times$ number of years (Objective 10.3A)

Compound Interest Earned

$CI = FV - PV$,
where CI is the compound interest (Objective 10.3A)

Present Value Formula for Use with Tables

$PV = FV \div$ compounding factor (Objective 10.3B)

Present Value Formula

$PV = \dfrac{FV}{(1 + i)^n}$,

where $i = \dfrac{r}{c} = \dfrac{\text{annual interest rate}}{\text{compounding periods per year}}$ and

$n = ct =$ compounding periods per year $\times$ number of years (Objective 10.3B)

To calculate the effective interest rate:

Calculate the future value of $100 at the nominal rate, and then find the interest earned on the $100. (Objective 10.3C)

REVIEW / TEST

1. Find the simple interest on a 45-day loan of $3000 at an annual interest rate of 8.75%.

2. Find the simple interest on a 3-month simple interest loan of $2700 at an annual interest rate of 7%.

3. A 90-day note is dated March 14. Find the maturity date of the note.

4. A 4-month note is dated July 31. Find the maturity date of the note.

5. To build inventory for the winter ski season, a sports shop signs a $5000, 9.6% simple interest note that is due in 60 days. Find the maturity value of the note.

6. To purchase wood for framing a house, a building contractor signs a $25,000, 10% simple interest note that is due in 9 months. Find the maturity value of the note.

7. To purchase a lathe, the owner of a cabinet shop signs a 6-month, 9% simple discount note that has a face value of $8000. Find the proceeds of the note.

8. A 7% simple discount note has a face value of $3400 and is dated April 23. The maturity date of the note is June 7. Find the proceeds of the note.

1. _____

2. _____

3. _____

4. _____

5. _____

6. _____

7. _____

8. _____

9. An executive placed all of a $5000 bonus check into an account that earns 8% interest compounded daily. What is the future value of the investment in 4 years?

10. A child received a gift of $1000, which was placed in an account that earns 6% annual interest compounded quarterly. Find the amount that will be in the account after 15 years.

11. The Arcadia Company wants to have $40,000 in 2 years to purchase new equipment. How much money should be placed in an account that earns 8% annual interest compounded monthly to reach this goal?

12. An electrical contractor wishes to have $80,000 in 4 years in order to expand the company's building. How much money must be deposited now in an account that earns 9% annual interest compounded monthly to reach this goal?

13. An investment will have a future value of $10,000 in 3 years. Find the present value of the investment if the interest rate is 10% compounded semiannually.

14. A savings and loan association offers a certificate of deposit at an annual rate of 9% compounded daily. Find the effective rate of interest.

15. The federal government offers an investment that earns 11% annual interest compounded daily. Find the effective rate of interest.

9. _____

10. _____

11. _____

12. _____

13. _____

14. _____

15. _____

APPENDIX

Answers to Chapter 1

SECTION 1.1   *pages 3 - 4*

You Try It 1

thirty-six million four hundred sixty-two thousand seventy-five

You Try It 2

452,007

You Try It 3

370,000

You Try It 4

4000

EXERCISES   *pages 5 - 6*

1. eight hundred five   **3.** four hundred eighty-five   **5.** two thousand six hundred seventy-five   **7.** forty-two thousand nine hundred twenty-eight   **9.** eighty thousand one hundred six   **11.** three hundred fifty-six thousand nine hundred forty-three   **13.** three million six hundred ninety-seven thousand four hundred eighty-three   **15.** 85   **17.** 406   **19.** 3456   **21.** 52,148   **23.** 609,948   **25.** 4,003,002   **27.** 9,463,000   **29.** 930   **31.** 1400   **33.** 7000   **35.** 44,000   **37.** 254,000   **39.** 60,000   **41.** 930,000   **43.** 4,000,000   **45.** 5,569,000   **47.** 6,850,000   **49.** 39,876,000

SECTION 1.2   *pages 7 - 12*

You Try It 1

$$\begin{array}{r} {\scriptstyle 2} \\ 95 \\ 88 \\ +\ 67 \\ \hline 250 \end{array}$$

You Try It 2

$$\begin{array}{r} {\scriptstyle 1\,1\ \ 2\,1} \\ 392 \\ 4{,}079 \\ 89{,}035 \\ +\ 4{,}992 \\ \hline 98{,}498 \end{array}$$

You Try It 3

$$\begin{array}{r} {\scriptstyle 15} \\ {\scriptstyle 4\ \cancel{5}\ 12} \\ 5\,4,\cancel{5}\,\cancel{6}\,\cancel{2} \\ -\,1\,4,4\,8\,5 \\ \hline 4\,0,0\,7\,7 \end{array}$$
Check: $\begin{array}{r} 14{,}485 \\ +\ 40{,}077 \\ \hline 54{,}562 \end{array}$

You Try It 4

$$\begin{array}{r} {\scriptstyle 13} \\ {\scriptstyle 5\,\cancel{3}\ \ 9\ 9\ 13} \\ \cancel{6}\,\cancel{4},\cancel{0}\,\cancel{0}\,\cancel{3} \\ -\,5\,4,9\,3\,6 \\ \hline 9,0\,6\,7 \end{array}$$
Check: $\begin{array}{r} 54{,}936 \\ +\ 9{,}067 \\ \hline 64{,}003 \end{array}$

You Try It 5

Strategy

To find the total number of patients treated, add the number treated in January (376) to the number treated in February (449).

Solution

$$\begin{array}{r} 376 \\ +\ 449 \\ \hline 825 \end{array}$$

The total number of patients treated was 825.

You Try It 6

Strategy

To find the amount that remains to be paid, subtract the down payment ($875) from the cost ($3350).

Solution

$$\begin{array}{r} \$3350 \\ -\ \ \ 875 \\ \hline \$2475 \end{array}$$

$2475 remains to be paid.

You Try It 7

Strategy

To find the number of units that must be sold during March:

◆ Find the total number of units sold during the first two months of the year by adding the number of units sold in January (225) and the number sold in February (198).

◆ Subtract the sum from the quota (650).

Solution

$$
\begin{array}{r} 225 \\ +\,198 \\ \hline 423 \end{array}
\qquad
\begin{array}{r} 650 \\ -\,423 \\ \hline 227 \end{array}
$$

The sales representative must sell 227 units in March in order to meet the quota.

EXERCISES   pages 13 - 16

1. 729   **3.** 156,499   **5.** 1584   **7.** 106,317   **9.** 1342   **11.** 120,570   **13.** 14,302   **15.** 24,218
17. 11,974   **19.** 6087   **21.** 77,159   **23.** 10,417   **25.** 9707   **27.** 66   **29.** 5002   **31.** 9   **33.** 574
35. 1344   **37.** 628   **39.** 378   **41.** 3621   **43.** 3642   **45.** 66,463   **47.** 4077   **49.** 52,404
51. 4367   **53.** 47 days   **55.** 2450 sports jackets   **57.** $15,000   **59.** 1883 miles   **61.** 731 boxes
63. 309 pairs   **65.** 199,442   **67.** $645,000   **69.** 72 potential customers   **71.** 252 units

SECTION 1.3   *pages 17 - 24*

You Try It 1

$$
\begin{array}{r} 756 \\ \times\,305 \\ \hline 3780 \\ 226\,80 \\ \hline 230{,}580 \end{array}
$$

You Try It 2

$7^2 = 7 \cdot 7 = 49$

You Try It 3

$$
\begin{array}{r}
705 \\
9\overline{)\,6345} \\
-63 \\
\hline
04 \\
-0 \\
\hline
45 \\
-45 \\
\hline
0
\end{array}
$$

Check:
$705 \times 9 = 6345$

You Try It 4

$$
\begin{array}{r}
470 \text{ r}29 \\
39\overline{)\,18{,}359} \\
-15\,6 \\
\hline
2\,75 \\
-2\,73 \\
\hline
29 \\
-0 \\
\hline
29
\end{array}
$$

Check:
$(470 \times 39) + 29 =$
$18{,}330 + 29 =$
$18{,}359$

You Try It 5

$$5\overline{)216{,}848}\ \ \ \frac{421\ r33}{}$$

```
        421 r33
515) 216,848
   – 206 0
      10 84
    – 10 30
        548
      – 515
         33
```

Check:
$(421 \times 515) + 33 =$
$216{,}815 + 33 =$
$216{,}848$

You Try It 6

Strategy
To find the total cost, multiply the unit cost ($6) by the number of units (35).

Solution

```
    35
  ×  6
   210
```

The total cost is $210.

EXERCISES *pages 25 - 28*

1. 623   **3.** 2492   **5.** 5463   **7.** 337,771   **9.** 3135   **11.** 44,100   **13.** 541,164   **15.** 342,171
17. 176,305   **19.** 406,927   **21.** 180,621   **23.** 1,302,725   **25.** 8   **27.** 81   **29.** 0   **31.** 81
33. 100   **35.** 1000   **37.** 3209   **39.** 90 r2   **41.** 204 r3   **43.** 778 r2   **45.** 3825   **47.** 3 r15
49. 21 r36   **51.** 200 r21   **53.** 4483 r18   **55.** 15 r7   **57.** 50 r92   **59.** 40 r7   **61.** 22¢   **63.** $17
65. $145   **67.** $12   **69.** $2200   **71.** $1800   **73.** $178,500   **75.** $200   **77.** 66¢   **79.** $4525

CALCULATORS *pages 29 - 30*

1. 15,000; 15,040   **2.** 7000; 7158   **3.** 2000; 2136   **4.** 7000; 6736   **5.** 1,200,000; 1,244,653   **6.** 540,000;
550,935   **7.** 800; 776 r12   **8.** 1000; 1072   **9.** 1,200,000; 1,208,917   **10.** 60,000; 61,800   **11.** 40,000;
38,283   **12.** 30,000; 33,573   **13.** 1,200,000; 1,138,134   **14.** 4,200,000; 4,315,403   **15.** 5000; 5129
16. 20,000; 21,967 r8   **17.** 6,300,000; 6,491,166   **18.** 600,000; 612,792   **19.** 1500; 1516 r170   **20.** 3000;
2886   **21.** 100,000; 125,665   **22.** 870,000; 845,181   **23.** 1300; 1296   **24.** 1700; 1665   **25.** $16,000
26. $14,000   **27.** 8000 cars   **28.** 20,000 miles   **29.** $20,000   **30.** an approximation

REVIEW/TEST *pages 33 - 34*

1. two hundred seven thousand sixty-eight (Objective 1.1A)   **2.** 1,204,006 (Objective 1.1A)   **3.** 75,000 (Objective
1.1B)   **4.** 96,798 (Objective 1.2A)   **5.** 135,915 (Objective 1.2A)   **6.** 100,332 (Objective 1.2A)   **7.** 9333
(Objective 1.2B)   **8.** 10,882 (Objective 1.2B)   **9.** 1685 (Objective 1.2B)   **10.** 726,104 (Objective 1.3A)
11. 6,854,144 (Objective 1.3A)   **12.** 28,000 (Objective 1.3A)   **13.** 64 (Objective 1.3A)   **14.** 703 (Objective
1.3B)   **15.** 8710 r2 (Objective 1.3B)   **16.** 1121 r27 (Objective 1.3B)   **17.** 1645 skateboards (Objective 1.2C)
18. 847 units (Objective 1.2D)   **19.** $17 (Objective 1.3C)   **20.** $2275 (Objective 1.3C)

Answers to Chapter 2

SECTION 2.1   *pages 37 - 40*

You Try It 1

```
     4
5) 22
  –20
    2
```
$\dfrac{22}{5} = 4\dfrac{2}{5}$

You Try It 2

```
     4
7) 28
  –28
    0
```
$\dfrac{28}{7} = 4$

You Try It 3

$14\frac{5}{8} = \frac{112 + 5}{8} = \frac{117}{8}$

You Try It 4

$45 \div 5 = 9 \qquad \frac{3 \cdot 9}{5 \cdot 9} = \frac{27}{45}$

$\frac{27}{45}$ is equivalent to $\frac{3}{5}$.

You Try It 5

Write 6 as $\frac{6}{1}$.

$18 \div 1 = 18 \qquad \frac{6 \cdot 18}{1 \cdot 18} = \frac{108}{18}$

$\frac{108}{18}$ is equivalent to 6.

You Try It 6

$\frac{16}{24} = \frac{16 \div 8}{24 \div 8} = \frac{2}{3}$

You Try It 7

$\frac{48}{36} = \frac{48 \div 12}{36 \div 12} = \frac{4}{3} = 1\frac{1}{3}$

EXERCISES   *pages 41 - 42*

1. $\frac{5}{4}$; $1\frac{1}{4}$ **3.** $2\frac{3}{4}$ **5.** 5   **7.** $2\frac{3}{10}$ **9.** $5\frac{1}{3}$ **11.** 16   **13.** $1\frac{15}{16}$ **15.** $\frac{7}{3}$ **17.** $\frac{13}{2}$ **19.** $\frac{59}{8}$

21. $\frac{21}{2}$ **23.** $\frac{34}{9}$ **25.** $\frac{11}{8}$ **27.** $\frac{4}{16}$ **29.** $\frac{21}{33}$ **31.** $\frac{125}{25}$ **33.** $\frac{44}{60}$ **35.** $\frac{35}{45}$ **37.** $\frac{60}{64}$ **39.** $\frac{1}{7}$

41. $1\frac{1}{9}$ **43.** $\frac{2}{15}$ **45.** $\frac{3}{5}$ **47.** $\frac{1}{8}$ **49.** $2\frac{1}{4}$

SECTION 2.2   *pages 43 - 50*

You Try It 1

$$
\begin{array}{r|ccc}
2 & 2 & 6 & 8 \\
3 & 1 & 3 & 4 \\
2 & 1 & 1 & 4 \\
2 & 1 & 1 & 2 \\
  & 1 & 1 & 1
\end{array}
$$

The LCM = $2 \cdot 3 \cdot 2 \cdot 2 = 24$

You Try It 2

$$
\begin{aligned}
\frac{7}{8} &= \frac{35}{40} \\
+ \frac{3}{5} &= \frac{24}{40} \\
\hline
\frac{59}{40} &= 1\frac{19}{40}
\end{aligned}
$$

You Try It 3

$$
\begin{aligned}
\frac{3}{4} &= \frac{30}{40} \\
\frac{4}{5} &= \frac{32}{40} \\
+ \frac{5}{8} &= \frac{25}{40} \\
\hline
\frac{87}{40} &= 2\frac{7}{40}
\end{aligned}
$$

You Try It 4

$$
\begin{aligned}
29 & \\
+ 17\frac{5}{12} & \\
\hline
46\frac{5}{12} &
\end{aligned}
$$

You Try It 5

$$7\frac{4}{5} = 7\frac{24}{30}$$
$$6\frac{7}{10} = 6\frac{21}{30}$$
$$+ 13\frac{11}{15} = 13\frac{22}{30}$$
$$26\frac{67}{30} = 28\frac{7}{30}$$

You Try It 6

$$\frac{13}{18} = \frac{52}{72}$$
$$- \frac{7}{24} = \frac{21}{72}$$
$$\frac{31}{72}$$

You Try It 7

$$17\frac{5}{9} = 17\frac{20}{36}$$
$$- 11\frac{5}{12} = 11\frac{15}{36}$$
$$6\frac{5}{36}$$

You Try It 8

$$8 = 7\frac{13}{13}$$
$$- 2\frac{4}{13} = 2\frac{4}{13}$$
$$5\frac{9}{13}$$

You Try It 9

$$21\frac{7}{9} = 21\frac{28}{36} = 20\frac{64}{36}$$
$$- 7\frac{11}{12} = 7\frac{33}{36} = 7\frac{33}{36}$$
$$13\frac{31}{36}$$

You Try It 10

Strategy

To find the market price of the stock at the end of the week, add the gain $\left(5\frac{3}{8}\right)$ to the market price at the beginning of the week $\left(37\frac{1}{4}\right)$.

Solution

$$37\frac{1}{4} = 37\frac{2}{8}$$
$$+ 5\frac{3}{8} = 5\frac{3}{8}$$
$$42\frac{5}{8}$$

The market price of the stock at the end of the week was $42\frac{5}{8}$.

EXERCISES   *pages 51 - 54*

1. 24   3. 24   5. 56   7. 20   9. 24   11. 180   13. $\frac{3}{7}$ 15. $1\frac{1}{6}$ 17. $\frac{17}{18}$ 19. $\frac{56}{57}$

21. $2\frac{1}{12}$ 23. $2\frac{17}{120}$ 25. $3\frac{2}{3}$ 27. $10\frac{1}{12}$ 29. $6\frac{7}{36}$ 31. $8\frac{1}{12}$ 33. $16\frac{5}{6}$ 35. $\frac{1}{3}$ 37. $\frac{19}{56}$

39. $\frac{4}{45}$ 41. $\frac{5}{48}$ 43. $\frac{17}{36}$ 45. $\frac{27}{70}$ 47. $\frac{23}{72}$ 49. $3\frac{1}{6}$ 51. $4\frac{1}{3}$ 53. $3\frac{2}{3}$ 55. $1\frac{3}{5}$ 57. $7\frac{5}{24}$

59. $8\frac{2}{5}$ 61. $4\frac{3}{8}$ 63. $10\frac{7}{8}$ 65. $22\frac{1}{8}$ 67. $\frac{7}{8}$ 69. $40\frac{3}{8}$ 71. $31\frac{1}{8}$ 73. $27\frac{7}{8}$ 75. $44\frac{3}{8}$

SECTION 2.3   *pages 55 - 58*

You Try It 1

$$\frac{4}{9} \times \frac{3}{10} = \frac{4 \cdot 3}{9 \cdot 10} = \frac{12}{90} = \frac{12 \div 6}{90 \div 6} = \frac{2}{15}$$

You Try It 2

$$3\frac{2}{7} \times 6 = \frac{23}{7} \times \frac{6}{1} = \frac{23 \cdot 6}{7 \cdot 1} = \frac{138}{7} = 19\frac{5}{7}$$

You Try It 3

$$\frac{3}{4} \div \frac{9}{10} = \frac{3}{4} \times \frac{10}{9} = \frac{3 \cdot 10}{4 \cdot 9} = \frac{30}{36} = \frac{30 \div 6}{36 \div 6} = \frac{5}{6}$$

You Try It 4

$$3\frac{2}{3} \div 2\frac{2}{5} = \frac{11}{3} \div \frac{12}{5} = \frac{11}{3} \times \frac{5}{12}$$
$$= \frac{11 \cdot 5}{3 \cdot 12} = \frac{55}{36} = 1\frac{19}{36}$$

You Try It 5

Strategy

To find the regular hourly wage:

◆ Find the overtime rate by dividing the overtime earned ($162) by the number of hours of overtime worked (9).

◆ Divide the overtime rate by $1\frac{1}{2}$.

Solution

$162 \div 9 = 18$

$$18 \div 1\frac{1}{2} = 18 \div \frac{3}{2} = 18 \times \frac{2}{3} = \frac{18}{1} \times \frac{2}{3} = 12$$

The regular hourly wage is $12.

EXERCISES   *pages 59 - 62*

1. $\frac{7}{12}$ **3.** $\frac{5}{12}$ **5.** $\frac{11}{20}$ **7.** $\frac{1}{8}$ **9.** $\frac{5}{12}$ **11.** $\frac{5}{14}$ **13.** 10   **15.** $\frac{4}{9}$ **17.** 10   **19.** $1\frac{4}{5}$ **21.** 3

23. $3\frac{4}{5}$ **25.** 19   **27.** $8\frac{2}{3}$ **29.** $3\frac{1}{2}$ **31.** $\frac{5}{6}$ **33.** 1   **35.** $\frac{7}{10}$ **37.** 3   **39.** $\frac{5}{6}$ **41.** $7\frac{1}{2}$

43. $\frac{1}{30}$ **45.** $1\frac{4}{5}$ **47.** $8\frac{8}{9}$ **49.** $\frac{1}{6}$ **51.** $1\frac{1}{11}$ **53.** $\frac{16}{21}$ **55.** $\frac{7}{11}$ **57.** $2\frac{13}{16}$ **59.** $1\frac{4}{15}$

61. $24 per hour   **63.** $135   **65.** $10 per hour   **67.** $8 per hour   **69.** $114   **71.** $525   **73.** $796
75. $623

CALCULATORS   *pages 63 - 64*

1. $7\frac{3}{4}$ hours   **2.** $7\frac{1}{4}$ hours   **3.** $7\frac{1}{4}$ hours   **4.** $7\frac{3}{4}$ hours   **5.** $7\frac{1}{2}$ hours   **6.** $37\frac{1}{2}$ hours   **7.** $7\frac{1}{2}$ hours

8. $7\frac{1}{2}$ hours   **9.** $6\frac{1}{4}$ hours   **10.** $6\frac{3}{4}$ hours   **11.** $6\frac{1}{4}$ hours   **12.** $34\frac{1}{4}$ hours

REVIEW/TEST   *pages 67 - 68*

1. $3\frac{3}{5}$ (Objective 2.1A)   **2.** $\frac{49}{5}$ (Objective 2.1A)   **3.** $\frac{20}{32}$ (Objective 2.1B)   **4.** $\frac{5}{8}$ (Objective 2.1C)   **5.** 60

(Objective 2.2A)   **6.** $1\frac{7}{12}$ (Objective 2.2B)   **7.** $1\frac{11}{18}$ (Objective 2.2B)   **8.** $7\frac{11}{12}$ (Objective 2.2C)   **9.** $\frac{1}{4}$

(Objective 2.2D)   **10.** $\frac{1}{8}$ (Objective 2.2D)   **11.** $13\frac{7}{8}$ (Objective 2.2E)   **12.** $\frac{18}{35}$ (Objective 2.3A)   **13.** $4\frac{2}{3}$

(Objective 2.3B)   **14.** $1\frac{13}{27}$ (Objective 2.3C)   **15.** $\frac{5}{7}$ (Objective 2.3D)   **16.** $1\frac{3}{19}$ (Objective 2.3D)

17. $12\frac{5}{8}$ (Objective 2.2F)   **18.** $44\frac{1}{2}$ (Objective 2.2F)   **19.** $8 per hour (Objective 2.3E)   **20.** $561 (Objective 2.3E)

Answers to Chapter 3

SECTION 3.1   *pages 71 - 76*

You Try It 1

two hundred nine and five thousand eight hundred thirty-eight hundred-thousandths

You Try It 2

27,600,000,000

You Try It 3

42,000.207

You Try It 4

$$\begin{array}{r} 0.03294 \\ +\,0.765 \\ \hline 0.79794 \end{array}$$

You Try It 5

$$\begin{array}{r} {\scriptstyle 1\,2} \\ 4.62 \\ 27.9 \\ +\;\;0.62054 \\ \hline 33.14054 \end{array}$$

You Try It 6

$$\begin{array}{r} {\scriptstyle 1} \\ 6.05 \\ 12 \\ +\;0.374 \\ \hline 18.424 \end{array}$$

You Try It 7

$$\begin{array}{r} 3.2\;\;\text{million} \\ +\,5.45\;\text{million} \\ \hline 8.65\;\text{million} \end{array}$$

You Try It 8

$$\begin{array}{r} {\scriptstyle 1\,1} \\ {\scriptstyle 6\;\not 7\;\;9\;13} \\ \not 7\,\not 2.\not 0\not 3\,9 \\ -\;\;8.4\,7 \\ \hline 6\,3.5\,6\,9 \end{array}$$

Check: $\begin{array}{r} 8.47 \\ +\,63.569 \\ \hline 72.039 \end{array}$

You Try It 9

$$\begin{array}{r} {\scriptstyle 16} \\ {\scriptstyle 2\;\;\not 8\;9\;9\;10} \\ \not 3.\not 7\not 0\not 0\not 0 \\ -\,1.9\,7\,1\,5 \\ \hline 1.7\,2\,8\,5 \end{array}$$

Check: $\begin{array}{r} 1.9715 \\ +\,1.7285 \\ \hline 3.7000 \end{array}$

You Try It 10

$$\begin{array}{r} {\scriptstyle 14} \\ {\scriptstyle 2\;\not 4\;\;9\;10} \\ \not 3\,\not 5.\not 0\not 0\;\text{billion} \\ -\;\;7.6\,7\;\text{billion} \\ \hline 2\,7.3\,3\;\text{billion} \end{array}$$

You Try It 11

Strategy

- To find the total of the miscellaneous expenditures, add the numbers in the row entitled Miscellaneous.
- To find the total amount spent on October 6, add the numbers in the column headed 10/6.

Solution

$8.75 + 4.27 = 13.02$
$34.90 + 75.50 + 4.92 + 4.27 = 119.59$

The total of the miscellaneous expenditures was $13.02. The total amount spent on October 6 was $119.59.

EXERCISES   *pages 77 - 80*

1. twenty-seven hundredths   **3.** one and five thousandths   **5.** thirty-six and four tenths   **7.** six and three hundred twenty-four thousandths   **9.** one and one hundred-thousandth   **11.** 0.762   **13.** 8.0304
15. 304.07   **17.** 5.36   **19.** 362.048   **21.** 5,230,000   **23.** 7,900   **25.** 3.107   **27.** 19.561
29. 37.185   **31.** 40.9820   **33.** 764.667   **35.** 34.8779   **37.** 69.66   **39.** 9.4 million   **41.** 92.5 thousand
43. 0.355   **45.** 4.088   **47.** 74.716   **49.** 8.666   **51.** 31.334   **53.** 5.627   **55.** 206.857   **57.** 4.5795
59. 4.685   **61.** 14.77 million   **63.** 8.6 thousand   **65.** $91.35   **67.** $14.02   **69.** $104.80   **71.** $148.77
73. $108.04   **75.** $72.86   **77.** $87.30   **79.** $18.86   **81.** $62.40   **83.** $132.50   **85.** $149.42
87. $43.12   **89.** $119.53   **91.** $17.38   **93.** $78.80   **95.** $189.37   **97.** $118.22   **99.** $126.24

SECTION 3.2   *pages 81 - 88*

You Try It 1

4.35

You Try It 2

3.291

You Try It 3

$$
\begin{array}{r}
870 \\
\times 4.6 \\
\hline
522\,0 \\
3480 \\
\hline
4002.0
\end{array}
$$

You Try It 4

$$
\begin{array}{r}
0.28 \\
\times\ 0.7 \\
\hline
0.196
\end{array}
$$

$0.28 \times 0.7 \approx 0.20$

You Try It 5

$6.9 \times 1000 = 6900$

You Try It 6

$$
\begin{array}{r}
2.7 \\
0.052\,\overline{)\,0.140\,,4} \\
-104 \\
\hline
36\,4 \\
-36\,4 \\
\hline
0
\end{array}
$$

You Try It 7

$$
\begin{array}{r}
0.4873 \approx 0.487 \\
76\,\overline{)\,37.0420} \\
-30\,4 \\
\hline
6\,64 \\
-6\,08 \\
\hline
562 \\
-5\,32 \\
\hline
300 \\
-228
\end{array}
$$

You Try It 8

$$
\begin{array}{r}
72.73 \approx 72.7 \\
5.09\,\overline{)\,370.20\,,00} \\
-356\,3 \\
\hline
13\,90 \\
-10\,18 \\
\hline
3\,72\,0 \\
-3\,56\,3 \\
\hline
15\,70 \\
-15\,27
\end{array}
$$

You Try It 9

$42.93 \div 100 = 0.4293$

You Try It 10

$2\dfrac{3}{4} = \dfrac{11}{4}$
$$
\begin{array}{r}
2.75 \\
4\,\overline{)\,11.00}
\end{array}
$$

You Try It 11

$5.35 = 5\frac{35}{100} = 5\frac{7}{20}$

You Try It 12

Strategy

To calculate the extensions and the total:
- Multiply the unit price in the first row by the quantity in the first row.
- Multiply the unit price in the second row by the quantity in the second row.
- Add the two products.

Solution

$12 \times 4.25 = 51.00$
$16 \times 9.45 = 151.20$
$51 + 151.20 = 202.20$

The extensions are $51.00 and $151.20.
The total is $202.20.

EXERCISES   pages 89 - 92

1. 7.4   **3.** 89.2   **5.** 670.97   **7.** 1.039   **9.** 8.63   **11.** 1946.40   **13.** 6.93   **15.** 13.50   **17.** 0.1323
19. 0.0602   **21.** 3.787   **23.** 20.148   **25.** 0.015397   **27.** 53.9961   **29.** 7   **31.** 3.9   **33.** 85.2
35. 6493   **37.** 6.3   **39.** 0.6   **41.** 2.5   **43.** 1.1   **45.** 0.8   **47.** 8.1   **49.** 0.30   **51.** 0.08
53. 18.75   **55.** 42.40   **57.** 40.70   **59.** 82.55   **61.** 0.625   **63.** 0.667   **65.** 0.167   **67.** 0.417
69. 2.003   **71.** 0.375   **73.** $\frac{4}{5}$ **75.** $\frac{8}{25}$ **77.** $\frac{1}{8}$ **79.** $1\frac{1}{4}$ **81.** $\frac{9}{200}$ **83.** $\frac{1}{3}$ **85.** $523.50
87. $74.25   **89.** $871.50   **91.** $470.25   **93.** $136.95   **95.** $2999.00   **97.** $373.75   **99.** $79.75

SECTION 3.3   pages 93 - 96

You Try It 1

3.07 m = 307 cm

You Try It 2

42.3 mg = 0.0423 g

You Try It 3

2 kl = 2000 L

You Try It 4

5682 m = 5.682 km

You Try It 5

Strategy

To find the freight charges:

- Find the total weight by multiplying the number of items (180) by the weight of each item (22.5).
- Divide the total weight by 100 kg, because the charge is based on the number of 100-kg units shipped. Round the quotient up to the nearest whole number.
- Multiply the number of 100-kg units shipped by the charge per 100 kg ($13).

Solution

$180 \times 22.5 = 4050$

$4050 \div 100 = 40.5 \approx 41$

$13 \times 41 = 533$

The freight charges are $533.

EXERCISES   *pages 97 - 98*

1. 420 mm   **3.** 4.200 L   **5.** 0.127 g   **7.** 6.804 km   **9.** 42.3 cm   **11.** 450 mg   **13.** 4.32 m
15. 1370 g   **17.** 88 cm   **19.** 45.6 mg   **21.** 4620 L   **23.** 2500 m   **25.** 37 ml   **27.** $1375
29. $187,000   **31.** $1087.20   **33.** 11,158 kg   **35.** $60   **37.** $2064   **39.** $142.50

CALCULATORS   *pages 99 - 100*

1. 55.855426688   **2.** 749.338785   **3.** 0.009765625   **4.** 0.001171875   **5.** $2099.87   **6.** $2568.41
7. $2112.10   **8.** $2686.49   **9.** $2027.30   **10.** $1917.84   **11.** $2457.50   **12.** $3989.11   **13.** $3727.65
14. $4847.35   **15.** $3305.40   **16.** $15,869.51   **17.** $207.81   **18.** $214.42   **19.** $331.49   **20.** $222.25
21. $269.44   **22.** $212.15   **23.** $352.47   **24.** $364.49   **25.** $316.14   **26.** $308.83   **27.** $391.82
28. $428.75   **29.** $1810.03

REVIEW/TEST   *pages 103 - 104*

1. forty-five and three hundred two ten-thousandths (Objective 3.1A)   **2.** 209.07086 (Objective 3.1A)
3. 11,800,000 (Objective 3.1A)   **4.** 458.581 (Objective 3.1B)   **5.** 27.76626 (Objective 3.1B)   **6.** 13.95 million (Objective 3.1B)   **7.** 0.074 (Objective 3.2A)   **8.** 2.4723 (Objective 3.2B)   **9.** 58,900 (Objective 3.2B)   **10.** 1.781 (Objective 3.2C)   **11.** 6.1924 (Objective 3.2C)   **12.** 0.8 (Objective 3.2D)   **13.** $\frac{7}{8}$ (Objective 3.2D)
14. $6\frac{2}{5}$ (Objective 3.2D)   **15.** 0.730 kg (Objective 3.3A)   **16.** 8.20 m (Objective 3.3A)   **17.** $273 (Objective 3.3B)   **18.** $32.76 (Objective 3.1C)   **19.** $0 (Objective 3.1C)   **20.** $8.10 (Objective 3.1C)   **21.** $0 (Objective 3.1C)   **22.** $8.25 (Objective 3.1C)   **23.** $6.01 (Objective 3.1C)   **24.** $15.63 (Objective 3.1C)   **25.** $17.64 (Objective 3.1C)   **26.** $0 (Objective 3.1C)   **27.** $7.79 (Objective 3.1C)   **28.** $14.06 (Objective 3.1C)
29. $55.12 (Objective 3.1C)   **30.** $449.85 (Objective 3.2E)   **31.** $249.75 (Objective 3.2E)   **32.** $499.90 (Objective 3.2E)   **33.** $1199.50 (Objective 3.2E)

Answers to Chapter 4

SECTION 4.1   *pages 107 - 114*

You Try It 1

East Phoenix Rental Equipment
3011 N.W. Ventura Street
Phoenix, Arizona 85280

NO. 2029

$\frac{68 - 461}{1052}$

September 30, 19 *94*

PAY TO THE
ORDER OF ___ *The First National Bank* ___ $ *694 $\frac{80}{100}$*

Six Hundred Ninety-Four and $\frac{80}{100}$ _____ DOLLARS

Meyers' National Bank
11 N.W. Nova Street
Phoenix, Arizona 85215

Eugene L. Madison

Memo _____
I: 1052 ''' 0461 I: 5008 2029 '''

You Try It 2

East Phoenix
Rental Equipment

DEPOSIT TICKET

DATE *October 4*, 19 *94*

MEYERS' NATIONAL BANK
11 N.W. Nova Street
Phoenix, Arizona 85215

I: 1052 ''' 0461 I: 5008 287 '''

ITEMS CREDITED SUBJECT TO VERIFICATION AND DEPOSIT AGREEMENT OF THIS BANK

CURRENCY	220	00
COIN		
C **LIST SINGLY** *68-837*	335	00
H *66-164*	74	28
E		
C		
K		
S		
TOTAL	629	28

BE SURE
EACH ITEM
IS PROPERLY
ENDORSED

You Try It 3

Strategy

To find the checkbook balance:
◆ Subtract the amount of each check from the previous balance.
◆ Add the amount of the deposit.

Solution

```
  785.93
− 189.43
  596.50
− 352.68
  243.82
+ 250.00
  493.82
```

The checkbook balance is $493.82.

You Try It 4

Strategy

To calculate the balance carried forward:
◆ Add the deposit to the balance brought forward.
◆ Subtract the amount of each check written.

Solution

```
  773.28
+ 294.63
 1067.91
− 146.50
  921.41
```

The balance carried forward is $921.41.

EXERCISES   *pages 115 - 116*

1.

```
East Phoenix Rental Equipment                          NO. 2847
3011 N.W. Ventura Street                                 68 - 461
Phoenix, Arizona 85280                                      1052

                                         January 24 , 19 95

PAY TO THE
ORDER OF    Xerox Corporation                    $  145 90/100

One Hundred Forty-Five and 90/100 _____ DOLLARS

Meyers' National Bank
11 N.W. Nova Street
Phoenix, Arizona 85215

Memo _____              Gloria B. Masters
I: 1052 III 0461 I: 5008  2847 II•
```

3. $1182.33   **5.** $247.63   **7.** $550.00   **9.** $678.49   **11.** $289.57   **13.** $461.06   **15.** $43.92
17. $75.00   **19.** $3427.01

SECTION 4.2   *pages 117 - 120*

You Try It 1

```
$2157.93
  324.66
   49.81
+ 155.79
$2688.19
− 838.45
$1849.74
```

You Try It 2

1. List the current checkbook balance:

2. Add checks outstanding and any additions to the balance:

3. Subtract deposits in transit and any bank charges:

		8105.79
Check No.	1769	216.84
Check No.	1770	938.36
Subtotal:		9260.99
		416.59
Subtotal:		8844.40
		5.50
Balance:		8838.90

4. The balance agrees with the balance listed in the bank statement.

EXERCISES   *pages 121 - 122*

1. $304.30   **3.** $526.04   **5.** $3250.46   **7.** $4308.38   **9.** $2328.03   **11.** 899; $192.14   **13.** $408.25
15. $4.75

CALCULATORS   *page 123*

1. the zero   **2.** none   **3.** the zero   **4.** both zeros   **5.** none   **6.** the last two zeros   **7.** the last two zeros

REVIEW/TEST   *pages 125 - 126*

1.

East Phoenix Rental Equipment
3011 N.W. Ventura Street
Phoenix, Arizona 85280

NO. 2936

$\frac{68 - 461}{1052}$

February 8 , 19 **95**

PAY TO THE
ORDER OF ___ *White River Company* ___ $ ___ *342 $\frac{97}{100}$*

Three Hundred Forty-Two and $\frac{97}{100}$ ———————————— DOLLARS

Meyers' National Bank
11 N.W. Nova Street
Phoenix, Arizona 85215

Memo _____

I: 1052 ''' 0461 I: 5008 2936 ''''

Gloria B. Masters

(Objective 4.1A)

2.

	East Phoenix Rental Equipment	CURRENCY	150	00
		COIN		
	DATE *Feb. 10* , 19 *95*	CHECKS LIST SINGLY 62-111	247	50
		66-722	199	98
DEPOSIT TICKET	MEYERS' NATIONAL BANK 11 N.W. Nova Street Phoenix, Arizona 85212			BE SURE EACH ITEM IS PROPERLY ENDORSED
		TOTAL	597	48

I:1052 '''0461 I:5008 287 ''•

ITEMS CREDITED SUBJECT TO VERIFICATION AND DEPOSIT AGREEMENT OF THIS BANK

(Objective 4.1A)

3. $796.10 (Objective 4.1B)   **4.** $1646.85 (Objective 4.1B)   **5.** $663.86 (Objective 4.1B)   **6.** $977.85 (Objective 4.1B)   **7.** $351.08 (Objective 4.1B)   **8.** $1298.70 (Objective 4.1B)   **9.** $593.47 (Objective 4.1B) **10.** $705.23 (Objective 4.1B)   **11.** $260.89 (Objective 4.1B)   **12.** $966.12 (Objective 4.1B)   **13.** $150.00 (Objective 4.1B)   **14.** $816.12 (Objective 4.1B)   **15.** $1275.94 (Objective 4.2A)   **16.** $2309.73 (Objective 4.2A)

Answers to Chapter 5

SECTION 5.1   *pages 129 - 134*

You Try It 1

$$\frac{4.6 = D - 5.8}{4.6 \mid 10.4 - 5.8}$$
$$4.6 = 4.6$$

Yes, 10.4 is a solution of the equation $4.6 = D - 5.8$.

You Try It 2

$$\frac{10p = 6}{10 \cdot \frac{4}{5} \mid 6}$$
$$8 \neq 6$$

No, $\frac{4}{5}$ is not a solution of the equation $10p = 6$.

You Try It 3

$$5.73 = A - 9.46$$
$$5.73 + 9.46 = A - 9.46 + 9.46$$
$$15.19 = A$$

The solution is 15.19.

You Try It 4

$$W(0.25) = 75$$
$$\frac{W(0.25)}{0.25} = \frac{75}{0.25}$$
$$W = 300$$

The solution is 300.

EXERCISES   *pages 135 - 136*

1. yes   **3.** no   **5.** yes   **7.** 2   **9.** 15   **11.** 6   **13.** 6   **15.** 8   **17.** 3   **19.** 5   **21.** 12.62 **23.** 23.9   **25.** 3.7   **27.** 12.4   **29.** $\frac{1}{2}$ **31.** 20   **33.** 360   **35.** 576   **37.** 109.88   **39.** 160 **41.** 25   **43.** 512.16